James E. Horne, PhD
Maura McDermott

The Next Green Revolution
Essential Steps to a Healthy, Sustainable Agriculture

Pre-publication
REVIEWS,
COMMENTARIES,
EVALUATIONS . . .

"Modern industrial agriculture, authors Horne and McDermott tell us, is deservedly on trial. Although it has given us an abundance of food, in the process it continues to cause serious environmental, health, and safety problems that endanger our natural resources, future food supplies, and the well-being of farmers, rural communities, and eventually all citizens. The remedy, their book describes with conviction and clarity, is what people now call a 'sustainable agriculture.' Written in an honest, down-to-earth style, this highly readable book gives us an insightful account of the struggle now underway to make sustainable agriculture a truly viable alternative to conventional farming. The authors' conviction is understandable and contagious."

Neill Schaller, PhD
Former Associate Director,
Henry A. Wallace Institute
for Alternative Agriculture,
Greenbelt, MD

More pre-publication
REVIEWS, COMMENTARIES, EVALUATIONS . . .

"What kind of agriculture do we need, and how can farmers provide it? These are the two questions addressed in Horne and McDermott's book, *The Next Green Revolution*. The book begins with an indictment of our current industrial agriculture for failing to fulfill its fundamental responsibilities to the farmers who use it, to the natural environment that supports it, and to the society that depends upon it. While their case against industrial agriculture is convincing, this book is really about the remedy, a treatment that can cure the ills of industrialization—a sustainable agriculture.

The authors refer to sustainable agriculture as revolutionary thinking, which it truly is. They suggest that we can and must find ways to meet our needs while leaving equal or better opportunities for others, both of this generation and for all generations in the future. Pursuit of individual short-run self-interests will not protect the natural environment or ensure long-run societal well-being. We must make conscious, purposeful decisions to take care of other people, the natural environment, and ourselves as well.

The bulk of the book is devoted to eight practical steps that farmers can take to ensure a healthy, enduring agriculture. Perhaps the greatest contribution of this book is its down-to-earth, step-by-step approach to developing more sustainable farming systems. It gives farmers practical suggestions for increasing profits and reducing risks while regenerating the soil, protecting the environment, and being good neighbors. Horne and McDermott show us that sustainable farming is not only possible, it is also very practical."

John E. Ikerd, PhD
Professor Emeritus,
University of Missouri,
Columbia

"James Horne tells his own story of how a sharecropper's son became a PhD agricultural economist and how his work with farmers led him to become a teacher, practitioner, and advocate in the search for more sustainable agricultural systems. By telling his personal story and vision, sandwiched with layers of practical information, Horne breathes life into the story of American agriculture, its history, triumphs, and tragedies.

Horne finds the full-time, small farmer—Jefferson's ideal citizen—as rare today 'as a Cadillac on an Oklahoma country road.' He builds a convincing case to show what America is losing by ignoring the social, economic, and ecological costs of these long-term and current trends in agriculture."

Lorraine Stuart Merrill, BS
Farmer and Agricultural Journalist,
Stratham, NH

More pre-publication
REVIEWS, COMMENTARIES, EVALUATIONS . . .

"*The Next Green Revolution* by Jim Horne and Maura McDermott is a very timely, informative, and readable contribution to the future of agriculture and rural communities in the United States. This book speaks to mainstream farmers in a very effective manner. It offers a vision of a more successful agriculture that supports both farmers and rural communities, and an eight-step plan for achieving it.

The heart of the book is the eight chapters that spell out the eight steps, from conserving and creating healthy soil (step 1) to increasing profitability and reducing risk (step 8). Each chapter covers, in a very informative and engaging manner, the principles behind the recommended step (both scientific and practical) and the basics of implementation, including handy 'how-to' checklists for farmers. The specifics of implementation have to be tailored, of course, to the individual farm, which will require considerable on-farm experimentation.

This book provides an excellent and very accessible starting point and guidebook for any farmer who is considering making changes. It is richly illustrated throughout with first-person examples from Horne's lifetime of experience in farming and consulting on both conventional and alternative methods, and with scientific findings from the USDA and other sources.

In short, the book is a very readable overview and blueprint for farmers who want to improve both profits and stewardship, but should also be read by researchers, policymakers, and anyone who has an interest in the future of agriculture or rural America."

Jill Shore Auburn, PhD
Former Associate Director,
University of California
Sustainable Agriculture Program;
Currently with USDA

Food Products Press®
An Imprint of The Haworth Press, Inc.
New York • London • Oxford

NOTES FOR PROFESSIONAL LIBRARIANS AND LIBRARY USERS

This is an original book title published by Food Products Press®, an imprint of The Haworth Press, Inc. Unless otherwise noted in specific chapters with attribution, materials in this book have not been previously published elsewhere in any format or language.

CONSERVATION AND PRESERVATION NOTES

All books published by The Haworth Press, Inc. and its imprints are printed on certified pH neutral, acid free book grade paper. This paper meets the minimum requirements of American National Standard for Information Sciences-Permanence of Paper for Printed Material, ANSI Z39.48-1984.

The Next Green Revolution
Essential Steps to a Healthy, Sustainable Agriculture

FOOD PRODUCTS PRESS
Sustainable Food, Fiber, and Forestry Systems
Raymond P. Poincelot, PhD
Senior Editor

Biodiversity and Pest Management in Agroecosystems by Miguel A. Altieri

The Next Green Revolution: Essential Steps to a Healthy, Sustainable Agriculture by James E. Horne and Maura McDermott

The Next Green Revolution
Essential Steps to a Healthy, Sustainable Agriculture

James E. Horne, PhD
Maura McDermott

Food Products Press®
An Imprint of The Haworth Press, Inc.
New York • London • Oxford

Published by

Food Products Press®, an imprint of The Haworth Press, Inc., 10 Alice Street, Binghamton, New York 13904-1580.

Cover design by Jennifer M. Gaska.

Library of Congress Cataloging-in-Publication Data

Horne, James E.
The next green revolution : essential steps to a healthy, sustainable agriculture / James E. Horne, Maura McDermott.
p. cm.
Includes bibliographical references (p.) and index.
ISBN 1-56022-885-7—ISBN 1-56022-886-5 (pbk.)
1. Sustainable agriculture—United States. I. McDermott, Maura. II. Title.

S441 .H67 2001
333.76'16'0973—dc21

00-064366

To my mother, Eva, my brother, Johnny,
and to the memory of my father, Earl Horne,
and to the fond memories of life on our farm.

ABOUT THE AUTHORS

In a day when specialization is king, **James E. Horne, PhD,** President of the Kerr Center for Sustainable Agriculture in Poteau, Oklahoma, is there to remind us that a holistic perspective can often provide the best answers to contemporary agriculture issues. Dr. Horne has a rare blend of diverse experiences in farming, policy, economics, education, and international agriculture that enables him to understand issues from various perspectives.

Dr. Horne was in the sustainable camp before it had recognition. He was among a handful who provided congressional testimony on the need for the USDA to initiate programs in sustainable agriculture. Once the USDA Sustainable Agriculture Research and Education program was established, he continued with it, providing leadership and direction. At the same time he was breaking ground at the Kerr Center, implementing over time a broad portfolio of programs in farm stewardship and farmer grants as well as developing the center's public policy and educational programs. He is also much in demand as a speaker.

Maura McDermott is Communications Director at the Kerr Center for Sustainable Agriculture in Poteau, Oklahoma. She is also a nationally recognized magazine writer, honored for her natural history/ environmental features in the state magazine *Oklahoma Today*. Additionally, she edits the Kerr Center's newsletter *Field Notes,* which goes to five thousand subscribers, and writes about farmers and agricultural issues. She specializes in making agricultural and environmental topics accessible and interesting to a broad range of readers.

CONTENTS

Foreword

The Next Green Revolution tells a story of change—of changes in American agriculture and of changes in ways of thinking about agriculture. Jim Horne's ways of thinking and of living clearly have been transformed as he has guided the Kerr Foundation out of the mainstream of agricultural respectability and into the less-traveled stream of agricultural sustainability.

I'm sure he asked me to write this foreword because I experienced a similar transformation. I first met Jim in the mid-1970s, shortly after I came to Oklahoma State University as an agricultural economist specializing in livestock marketing. At the time, Jim and I were both fairly traditional and conventional in our thinking about economics and agriculture. We truly believed that the markets worked—that the greatest good for society resulted from everyone pursuing their own self-interests through buying and selling in open markets. We depended on Adam Smith's "invisible hand" to transform individual greed into societal good. Farming had to be a bottom-line business, first and foremost; everything else was irrelevant if a farm wasn't profitable. If individual farmers took care of their own business, then the business of agriculture would take care of itself. We thought like "good economists."

However, we both began to question the conventional wisdom of traditional agriculture during the farm financial crisis of the 1980s. Something clearly wasn't working and it wasn't all the farmers' fault. Certainly, many had borrowed too much money during the boom times of the 1970s, but that's what we experts had encouraged them to do. Booming export markets were supposed to more than keep pace with any feasible expansion in U.S. production. The good times were supposed to last for a long time, if not forever. But they didn't. By the mid-1980s, American agriculture was in crisis. Export markets dried up and prices of farm commodities tumbled. Farms that had survived for generations failed; families that had endured for decades experienced depression and divorce; rural communities that once had prospered withered and died. Something was fundamentally wrong.

As economists, both Jim and I came to the sustainable agriculture movement searching for a way to sustain farms economically. We realized that farmers had been made increasingly dependent on purchased inputs, particularly fertilizers and pesticides. We also knew that while prices of farm commodities had gone down as well as up, prices of purchased inputs had continually climbed. We were looking for a way to help farmers escape the tightening grip of rising costs and fluctuating prices.

It was only after we began to understand and reject the conventional, industrial agricultural paradigm that we could begin to see that the economics of agriculture were inextricably linked with its ecological and social foundation. Farms quite simply cannot be sustained economically unless they are ecologically sound and socially responsible. A farm that fails to protect and conserve the resources that support its productivity eventually loses its ability to produce and, therefore, cannot remain profitable. A farm that fails to support the community and society for and in which it exists fails in its fundamental purpose and thus will fail economically. However, it should be pointed out that farms cannot be ecologically sound or socially responsible unless they are also economically viable—all three facets are necessary and none alone is sufficient. But the bottom line is that economists and farmers alike must come to realize that the long-run economic viability of any farm rests upon its ecological and social foundation.

This book begins with an indictment of industrial agriculture for failing to fulfill its fundamental responsibilities to the farmers who use it, to the natural environment that supports it, and to the society that depends upon it. The indictment accuses industrial agriculture of jeopardizing the inheritance of our children, as well as agriculture's future productivity, by endangering essential natural resources. It accuses industrial agriculture of peddling addictive agricultural chemicals to farmers for profit while ignoring the environmental and human health consequences. Finally, industrial agriculture is indicted for bankrupting farmers, destroying rural communities, and leaving rural America open to exploitation. Following the indictments is the development of the case against industrial agriculture. Although the case is convincing, this book is really about the remedy, a treatment that can cure the ills of industrialization: a sustainable agriculture.

Jim uses his personal involvement with USDA programs in the early days of the sustainability movement as a backdrop for explaining what sustainable agriculture is all about. He refers to sustainable agriculture as revolutionary thinking, which it truly is. It's based on a revolutionary worldview—the world is not like a big complex machine; it is like a big complex living organism. Pursuit of individual short-run self-interests will not protect the natural environment or ensure long-run societal well-being. We must make conscious, purposeful decisions to take care of other people and to care for the natural environment, as well as to take care of ourselves. Humanity will not sustain itself automatically. We can and must find ways to meet our needs while leaving equal or better opportunities for others, both of this generation and for all generations in the future. We didn't know it at the time, but the USDA Sustainable Agriculture Research and Education program probably marked the historical beginning of a new green revolution.

Definitions are necessary for communication, but they don't necessarily cause anything positive to happen on the land. So the rest of the book focuses on proposed actions—the eight "essential steps to a healthy, sustainable agriculture."

Step 1 is to create and conserve healthy soil. All of life arises from the soil. Certainly, living things need air, water, and sunlight to survive and grow, but they also must have soil. Animals eat plants that grow from the soil, and even sea dwellers must have minerals dissolved from the soil. Soil conservation has been the focus of past efforts to protect the soil. However, sustaining productive soil requires much more than simply keeping it from washing or blowing away. A healthy soil is a living organism. Soil breathes air, drinks water, stores energy, and is filled with living creatures. A healthy soil must be nurtured and cared for—it shouldn't be treated like "dirt."

Step 2 is to conserve water and protect its quality. Water, like soil, is essential for life. In fact, most living things, including the human body, are made up mostly of water. But just any old water won't do. Many things dissolve in water; that's one of its most important properties. But many of those things that dissolve in water, including salt, make it unfit for use by humans and other living beings. As Jim and Maura point out, only about 3 percent of all water is "fresh" water, much of which is contained in the polar ice caps. Fresh water is created for the most part by evaporation from the oceans. Water vapor in the clouds falls back to earth as rain and eventually makes its way back to the sea. Human life depends on the quantity and quality of these fragile streams that begin with rainfall and end in the oceans. To sustain life, we must conserve and protect these streams of fresh water.

Step 3 is to manage organic wastes to avoid pollution. Agricultural wastes and chemicals have become one of the most significant sources of water pollution in the United States. Pollution arising from manufacturing sources was addressed first because it was the most obvious and the easiest to address. However, as pollution from manufacturing sources was reduced, continued industrialization made agriculture an increasingly important polluter of the environment. Pollution is an inherent consequence of industrialization. Industrialization requires that production processes be concentrated in one place—whether in an automobile assembly plant or a large-scale confinement animal feeding operation. Industrialization requires control—whether by standardization of manufacturing processes or through use of pesticides and fertilizers to ensure predictable crop performance. Pollution is an inherent consequence of the concentration and standardization of production processes. The only real solution to pollution is to abandon the industrial model of production—particularly in agriculture.

Step 4 is to select plants and animals adapted to the environment. A fundamental principle of sustainability is to work in harmony with the natural environment. Industrialization attempts to conquer nature; sustainability requires that we work with nature instead. Hunters and gatherers simply selected from whatever grew, as nature collected solar energy and transformed it into plants and animals. Agriculture attempts to tip the ecological balance in favor of humans relative to other species—to capture more solar energy for human use. But sustainability dictates that we not tip the balance too far, or we risk destroying the integrity of the natural system, of which we also are a part. Over time, plants and animals have naturally evolved and adapted to specific climates and to physical and biological environments. Given time, living things will find their "place in the sun." Sustainable agriculture attempts to work with nature by putting plants and animals in the environments where they grow best with the least help. Farmers can then focus on increasing production and reducing costs by caring for and nurturing natural processes, rather than creating artificial environments and trying to force growth.

Step 5 is to encourage biodiversity. Nature is inherently diverse. Industrialization requires specialization and thus is inherently in conflict with nature. Sustainability seeks harmony with nature through diversity. The productivity of nature arises from positive interrelationships. Bacteria and other microorganisms feed on dead plants and animals and, in the process, provide food for other living plants and animals. Animals feed on plants, but plants also feed on wastes from animals. People feed on plants and animals and, in turn, provide food and care for plants and animals. All of these living processes are interrelated in critical ways with the nonliving elements of the environment—the soil, water, and air. All living things, including humans, are part of an incomprehensibly complex web of life. Each node in this web performs a unique and different function in support of the rest of the web. Thus, each element—each microorganism, each plant, each animal—plays a potentially important role in maintaining the health and well-being of the web as a whole. Farms that work in harmony with nature must be likewise diverse. The productivity that arises in nature comes from positive interrelationships among microorganisms, plants, and animals; among economic enterprises; and among the people who work on a farm.

Step 6 is to manage pests for minimal environmental impact. Perhaps the greatest challenge to ecologically sound farming is to manage pests without damaging the natural environment. We now know, with relative certainty, that the use of many agricultural chemical pesticides represents a significant risk of harming the natural environment as well as a risk to human health. But pests are difficult to control without these chemicals. Nature attempts to keep things in balance—there are natural controls and deterrents that prevent any one species from maintaining a dominant position in nature. The things that we call pests—insects, weeds, diseases, etc.—are simply nature's means of keeping in check the crop or livestock species that we want to grow more of than nature would provide on its own. So agriculture is an inevitable imposition on nature; but ways exist to minimize this imposition and thus to control pests with a minimal environmental impact. All pests have natural checks and balances within nature as well. When we maximize use of these natural and beneficial checks and balances, we minimize the need for synthetic chemical pesticides and thus minimize the environmental impact. We must learn to be very careful when we interfere with nature.

Step 7 is to conserve nonrenewable energy resources. Many of the resources that support agriculture, and thus support human life on earth, are not renewable within any reasonable time span. Fossil fuels, for example, were formed over many millions of years; they presumably would take as many millions of years to renew. The foundation for industrial agriculture is fossil fuels. In fact, industrial agricultural production and distribution now uses up more fossil fuel energy than it captures in solar energy. Agricultural mechanization and food distribution are highly dependent on fossil fuels. Many agricultural pesticides and fertilizers are fossil fuel-based as well. A fossil fuel-based agriculture quite simply is not sustainable in the long run. This does not imply that agriculture should stop using nonrenewable resources; it simply means that we need to move toward farming systems that utilize regenerative, renewable resources as the remaining fossil fuel stocks inevitably continue to decline. Such systems exist; they just need much more research, development, and public support if they are to become sufficiently productive to support humanity after the fossil fuels are gone.

Step 8 is to increase profitability and decrease risk. Perhaps the greatest challenge in farming sustainably is to maintain profitability without exploiting either the natural environment or people. A farm that is not economically viable quite simply is not sustainable. However, the recent preoccupation with short-run economic thinking has created an agricultural system that seems to force farmers to exploit every profit opportunity to survive in a highly competitive market environment. Publicly held corporations are not people; they have neither heart nor soul. Thus, corporations have no true sense of friendship or stewardship—they create relationships and conserve resources only to add to the economic bottom line. Farmers in competition with corporations seem forced to exploit and pollute in order to succeed.

A better way exists. Thousands of farmers across the country are finding ways to succeed financially without exploiting nature or people. They are farming profitably, but they are not maximizing profits at the expense of their community or the environment. They are finding ways to bring the economic, ecological, and social dimensions of their farming operations into balance and harmony. They are reducing costs through low-input farming—by letting nature do more of the work. They are increasing value by niche marketing—by giving consumers what they really want and what they value individually. They are increasing overall productivity by utilizing their unique abilities and talents—by doing things that they do well and for which they have a personal passion. They are building unique relationships between themselves and the land and between themselves and their customers—they are linking people with purpose and place. Many of these farmers find that their profits actually increase as they focus less on profits and more on balance and harmony—as they focus on overall quality of life.

Jim and Maura don't dwell on the philosophical principles, as I have done here. They instead outline very practical examples of how farmers individually can implement each of the eight steps in developing more sustainable farming systems. Jim tells us how he learned lessons by relating them to his practical experiences as the director of the Kerr Center for Sustainable Agriculture, as a typical part-time family farmer, and as a real person on life's journey. With each step comes a checklist, so farmers will have a ready reference list to help guide them on their quest toward sustainability.

Farming in the United States today truly stands at a fork in the road to the future. However, the choice is no longer between conventional agriculture and sustainable agriculture. Agriculture as we have known it for the past several decades is coming to an end. The final stage of industrialization will mean the end of family farming and the beginning of, essentially, complete corporate control of the food and fiber system. The choice for the future is between a corporate, contract-based agriculture, driven solely by profit and growth, and an independent, family-based agriculture that balances the economic, social, and ecological dimensions of quality of life for long-run sustainability. Corporate contract agriculture may seem easy, but it quite simply is not sustainable for farmers or for society in general. Sustainable agriculture most certainly is not simple; in fact, it is downright difficult. But more farm families are choosing the road less traveled and finding that it makes a tremendous difference in the overall quality of their lives.

Neither Jim nor I are traditional or conventional in our thinking about economics and agriculture any longer. We have lived through a transformation. We no longer believe that the markets work very well. The greatest good for society is not served when everyone pursues their own self-interests. Adam Smith's "invisible hand" has been mangled in the machinery of industrialization. The markets are no longer competitive, at least in an economic sense. Farming must be more than just a bottom-line business. Everything else is *not* irrelevant. A farm must be ecologically sound and socially responsible or it cannot sustain its profitability.

Sustainability is not just about farming; it is about a better way of life. We must make conscious, purposeful decisions to take care of one another, to take care of the natural environment and to take care of ourselves. Anything less is a life out of balance—a life of conflict and disharmony. Anything less is a life without real quality. Anything less is not sustainable. The next green revolution must be sustainable.

John E. Ikerd, PhD
Professor Emeritus of Agriculture Economics
University of Missouri

Preface

What is a healthy, enduring agriculture? Once you have an idea of what it is, how do you practice it? And once you have some success at it, how do you convince others to change—to try something new?

These have been important questions for us at the Kerr Center for Sustainable Agriculture. Since 1985, when the Kerr Center came into being, answering these core questions has been our task. In this book, we attempt to answer the first two questions, and, in doing so, persuade others to try new approaches to agriculture.

A healthy, enduring agriculture is a sustainable agriculture, and this term has become the umbrella term for approaches to agriculture that are environmentally friendly, profitable, and fair to farmers and ranchers. This book largely grew out of Jim's experiences in Oklahoma at the Kerr Center, as well as work on the regional and national level with the USDA's Sustainable Agriculture Research and Education (SARE) program. We wrote this book to convince those people who are unfamiliar with, or perhaps suspicious of, sustainable agriculture that change is both needed and possible.

Although the situation has improved in recent years, there is still a lack of information about new approaches to farming and many misconceptions about alternative agricultural approaches. It is our intention to remedy that situation by writing an easy-to-read, practical introduction to the subject, with the goals of a healthy agriculture synthesized into eight comprehensive steps.

The time is right for this book; there is more interest in sustainable agriculture than ever before. Since 1985, the number of farmers and ranchers around the United States and Canada who are making their farms more sustainable has been slowly but steadily rising. These farmers are reasonable people who have recognized that to survive and prosper, they must try something new. There is nothing offensive in their approaches, although some proponents of conventional agriculture have attempted to brand these alternative farming methods as naive and unrealistic.

The eight steps presented in this book, too, are reasonable and pragmatic. The Kerr Center's approach to change has been to meet farmers where they are and then slowly move them toward sustainable farming. New farming systems don't happen overnight; they must evolve.

The eight steps are essential for a healthy, sustainable farming operation and are, by extension, essential principles for a sustainable agriculture. Two other principles are also key: A sustainable agriculture enhances the quality of life of farm families and revitalizes rural communities. Taking good care of both human and environmental resources is the heart of a sustainable agriculture. Unfortunately, the demands of conventional agriculture for farmers "to get big or get out," or, in more recent times, to become low-paid workers in vertically integrated corporate "operations," force agriculture in an unhealthy direction. Such trends erode the quality of farm life and the vitality of rural communities as well as threaten the health of the natural environment.

We decided to write this book in the first person (from Jim's point of view) in order to make it easy for readers to follow the journey of someone perhaps not very different from themselves. It is the story of someone who slowly came to grips with the failure of the conventional system of industrial agriculture and began searching for a better way. We hope presenting his individual journey of change will point the way for others who feel a similar dissatisfaction with the current state of affairs in agriculture.

This book is intended to be used for education and outreach in agriculture programs and courses. It is also for anyone who has an interest in the future of agriculture. Both ordinary citizens and those in positions of political power reading this book will realize that the current industrial agricultural system is broken. How to fix it? Too often, what are presented as "new" solutions are, in reality, just tired old policies, respun. Genuinely new approaches are needed. The root causes of the farm problem must be examined and solutions found that take care of nature, keep people on the farm, and provide meaningful employment in rural communities.

This book, although including much analysis of the harm that industrial agriculture has brought to rural America, is ultimately a positive book. Perhaps USDA programs should also focus on a positive rather than negative approach. Why not pay farmers to implement practices that would fulfill the eight steps rather than pay them later to clean up pollution or compensate for overproduction? We hope agricultural economists, in particular, will see the wisdom in this book. As a group, agricultural economists have made so many wrong assumptions about farmers and natural resources—that such things as lost topsoil, lost biodiversity, and lost rural communities should not be included when calculating the bottom line. Natural resources are not infinite, and healthy rural communities are important; such "intangibles" have value and should be counted.

Many people helped make this book possible. The founder of the Kerr Center, Kay Kerr Adair, not only launched Jim Horne's and the Kerr Center's quest for a more sustainable agriculture but offered much encouragement during the writing of the book. The entire Kerr Center board of trustees offered patient support. Trustee Lloyd Faulkner, in particular, encouraged the idea that this book be different and offer fresh insights into the subject. Staff and friends at the USDA's SARE program, both in the national office and on the administrative council of the southern region, encouraged the effort, as did Kerr Center staffers, past and present. Communications assistant Liz Speake was of great help in preparing the bibliography and completing many other tasks. Ken Williams, a former staff member and now aquaculturist at Langston University, was of particular importance in helping Jim formulate the key points of a sustainable system, the foundation upon which this book was built.

A special thank you is extended to our editor, Dr. Raymond Poincelot. His insights made this a better book.

Last but not least, we want to thank our families—Jim's wife, Brenda, and children, Doug and Andrea, and Maura's husband, Ron Wood—for supporting us in this effort, for recognizing the importance of a sustainable agriculture, and for allowing us to forfeit family time in order to write this book.

It is urgent that the message of sustainable agriculture be heard. Farmers, consumers, policymakers, economists, and agricultural educators and professionals must join together to stop farming communities from deteriorating further. If we don't do this now, future generations in rural America (and in urban America, too) will have even fewer choices and less freedom than we have today.

Chapter 1

On Trial: Industrial Agriculture

> As we are part of the land, you too are part of the land. This earth is precious to us. It is also precious to you . . .
>
> Chief Seattle
> Suquamish (Native American)

I grew up in the cotton fields, the son of a sharecropper in southwestern Oklahoma. Our lives revolved around the cotton plant. It wasn't the dreamy television cotton, but in a profound way it did form "the fabric of our lives." We sowed the cottonseed in May, chopped out the weeds during the dog days of summer, and pulled off the fluffy white bolls on brilliant fall days when everything in the flat country seemed to shimmer. I loved cotton, but probably if we had grown corn, I would have felt the same way about it. It was farming that I loved, and farming which formed my character. On our farm, I acquired stamina. I learned perseverance and felt pride in jobs well done. And it was there that I began to learn the lesson that I am still learning: The good earth will fail us if we fail her—but she will sustain us if we treat her right.

Farming is in my blood. I am fifty-three years old: I grew up on a farm and ever since graduating from college I have raised cattle. I have worked with farmers or studied the problems of agriculture my entire life. In doing so, I have been witness to a great drama—the story of farmers and farming in the United States in the latter half of the twentieth century.

Sad to say, until recently, it's been largely a tragedy. I have seen the farming life disappear piece by piece, with families bankrupted and displaced and rural communities turned into ghost towns—all the while watching the quality of our soil and water decline and the balance of nature upset, along with the towns and the lives of the people.

The good earth will fail us if we fail her—but she will sustain us if we treat her right.

Perhaps because it has been going on for fifty years, the story has lost its dramatic punch for the public in this age of short news bites. Besides an occasional news story during the occasional declared "farm crisis," the public remains largely uninformed about what I think is a major threat to all of us. That's why, if I could somehow miraculously be in charge of a major television network, I would have the farm story dominate the news the way the investigation and impeachment trial of President Clinton or the trials of O. J. Simpson did. It is a story much more complex than either of those stories and, ultimately, much more important.

But how to present it? Given our apparent love for a good courtroom drama, perhaps the tragedy of industrial agriculture would have to be presented as a trial. The opening day might go something like this:

It is a sunny Monday morning in March. The place: a courthouse in a rural town. The occasion: a trial that I hope will open some eyes, change some minds, and right some wrongs.

The hearing is being televised and is about to begin. A farmer walks to the front of the courtroom, takes his reading glasses out of his pocket, puts them on, and arranges the papers in his hand. The courtroom hushes. Behind him on the wall hang the words of Thomas Jefferson: "The small landholders are the most precious part of the state." The farmer is one of those small landholders, a middle-aged man wearing blue denim overalls and a farm cap bearing the name of his hometown grain elevator. He is not a particularly romantic figure—he has a belly under the overalls and he is red-faced from being in the sun a lot and he tends to be suspicious of urban people and urban style.

He himself is out of fashion, as are his problems. He struggles to keep farming because it is his life: He is bound to the soil by his own toil and by the toil of his ancestors, and by the hope that one day his children will be able to earn an honest living from that same soil.

The farmer visits the courthouse every year to pay his taxes. Built 100 years ago of native sandstone quarried in the hills nearby, it once sat in the center of a vast prairie. Now the courthouse sits in the midst of fields and pastures that stretch to the horizon.

During the past week, the farmer has noticed a faint wash of green spread across his pastures—new grass for his cows after a long winter of eating hay. The wild plums in the fence row are budding out. All over the county—by the courthouse steps and farmhouse doors—daffodils are blooming. It's not quite spring and not winter anymore either; it's a time of change. The growing season is poised to begin. But first there is some business to take care of.

The farmer begins to read the document in his hand:

> *Industrial agriculture—defined as the current predominant system of agricultural production and its supporting establishment—stands accused, in a three-part indictment, to wit:*
>
> *of endangering the essential natural resources of soil, water, and life, thereby jeopardizing the future productivity of agriculture and the inheritance of our children;*
>
> *of hooking farmers on fossil fuels, and the fertilizer and pesticides made from them, while downplaying the consequences of overusing such products;*
>
> *of desolating rural America by bankrupting farmers and ignoring the well-being of rural communities, thus leaving them open to exploitation.*
>
> *These crimes show a reckless disregard for the life and health of farmers, rural communities, and the natural world, jeopardizing our ability to feed an ever-growing population. As a result, the food security of our nation, and our world, is threatened.*

The words echo in the courtroom. Who are the defendants? They are the men and women who have bought the line that we must preserve the agricultural status quo or the world will starve; the people who believe a farm should be run like a factory and that bigger is better; the people who sidestep questions about the health of natural resources and the diminishing number of farmers. These people work for the corporate giants of agriculture, in the United States Department of Agriculture, and in the university-based research and education system. Of course, among them are farmers who buy their products, sign up for their programs, and listen to their advice. The institutions represented are entwined and have defined the agricultural debate for the past fifty years.

As the indictment is read, they look down at their shoes or up angrily at the farmer. Naturally, they don't appreciate anyone questioning what they see as the success and efficiency of industrial agriculture. Above them float ghosts or guardian angels, if you will—insiders, staunch pillars of the agriculture establishment such as former Secretaries of Agriculture Earl Butz and Ezra Taft Benson.

Over on the plaintiff side sits a smaller group: farmers and ranchers, along with researchers from universities and people from the Department of Agriculture, a few agribusinessmen, activists from nonprofit organizations, and consumers concerned about the safety of their food and how it is grown. Their expressions are a mix of sorrow and regret, mixed with a few satisfied smiles at having their grievances taken seriously at last. The plaintiffs believe in something called sustainable agriculture. They don't have much use for the status quo. They don't believe that bigger is necessarily better; they don't believe that a farm should be run like a factory, They do believe that the health of our natural resources and the environment, and the physical and financial health of farmers, are questions central to the future of agriculture. They too have their guardian angels: outsiders, thorns in the side of the establishment like conservationists Aldo Leopold and soil researcher Sir Albert Howard.

Glancing at both sides, there is not much that distinguishes them from one another, especially the farmers. Farming is not for the young anymore; most of them are over forty. Most are wearing jeans and boots. However, a few farmers stand out from the rest—Amish men, bearded, in black coats—keeping to themselves in the back of the courtroom. They are a link to our ancestors, those men with solemn faces that stare at us from ancient brownish photographs, for whom the dream of owning their own land steeled them to the worst hardships; men who sometimes died in their fields, behind a horse and plow.

The fact is, farmers are on both sides of the argument because American farmers have both contributed to the crimes of modern agriculture and been victimized by them.

Someone calls out, "Who is the judge?"

The farmer who read the indictment looks straight into the camera. "You are," he says to the world.

In this television courtroom, I sit on the plaintiffs' side, but I know the men on the other side, too, having spent many years among them. Although I don't wear overalls most days of the week, I am a part-time farmer like most of those who own small farms. My full-time work is being the leader of a private, nonprofit foundation in Oklahoma, the Kerr Center for Sustainable Agriculture. I have been part of this organization for more than twenty-five years, first as an agricultural economist and member of a consultation team that advised farmers in southeastern Oklahoma, and then as president. My organization and I have changed over the years—changed sides, in fact, a rarity in agriculture. Once I unquestioningly supported the industrial side. In this trial, I will be a witness for the plaintiffs and tell what I know.

I was born, like many of the farmers in my imaginary courtroom, at the beginning of the post–World War II agricultural revolution. It has indeed been a revolution, with all the attendant upheavals of any revolution. Although farmers have been most affected, everyone has been touched by it. I have been a firsthand witness to the changes wrought by it, starting out as a supporter, and then changing my mind. I offer myself as a kind of everyman—a man who has been both a part of events and an observer of what has taken place.

I grew up on a farm in the small community of Cold Springs in Kiowa County, Oklahoma. Until I was thirteen years old, my family were solely sharecroppers; then we bought a farm of forty acres. Each year, we generally leased 400 more acres to grow cotton, wheat, and barley. We often raised pigs, the poor man's livestock. The people in our community were real farming pioneers—what with boll weevils, hailstorms, and drought, it took brave-hearted people to farm on the southern Great Plains. But farm we did. Farming was everything; there were no factories or big cities nearby. And although no one would ever say this out loud, we accepted what the government, newspapers, and extension agents told us: that farming was a noble calling; the world depended on us, American farmers, to feed and clothe its burgeoning population.

Our farm had one foot in the past. We didn't get running water until 1960 when we bought our place, and we did some farm work the old-fashioned way: my mother and I chopped cotton (hoed out the weeds between the plants in the row) in the summer. Each long row was as straight as a section-line road: my father, I was told recently, planted the straightest rows in the county, a mark of distinction among farmers there. Hoeing that cotton, all 160 acres of it, in mile-long rows, was miserably hot work that had to be done daily if we were going to get through it. We even worked on the Fourth of July, quitting at noon if we were lucky. What made the work tolerable was talking to my mother about school, my teenage love life, and my future. I remember her telling me: "You need to get an office job, son, so you won't have to do this for the rest of your life." As a farmer's wife, she had sacrificed much and wanted better for me.

Despite the hard work, farming had a hold on me. For one thing, I loved to plow. I'd go out after supper—we had a television but I didn't watch it much—and I'd plow until midnight or one in the morning. I enjoyed being outside and smelling the fresh dirt as it was loosened. And I liked tractors; I could match any of my friends in tractor knowledge.

During the 1950s, secretary of agriculture Ezra Taft Benson had defined the future when he warned farmers "to get big or get out." In the early 1960s in Kiowa County, these words had been taken to heart: the bigger the tractor and plow, the more land you could farm, the more money you could make, and the more status you had. The future had arrived. On the school bus my friends and I would look at farming magazines and talk tractors—which was bigger, could pull more, or plow a wider swath of ground. We believed we had every reason to be proud; we were on the front lines of a great agricultural revolution, the likes of which the world had never seen. And it was just beginning. Crop and animal production in the United States was on its way up; by the early 1990s, production nearly doubled from the 1960s. How many people a lone American farmer could feed rose astronomically—from 10.7 people in 1940 to 25.8 in 1960 to 75.8 in 1970. In the 1990s, according to the USDA, an American farmer provided food and fiber for 129 people.

American consumers reaped the benefits of this productivity—the real price of farm commodities dropped during this time, and food remained cheap. In fact, the percentage of disposable income consumers spend for food has declined in the past twenty-five years. United States consumers spend a little over 10 percent, the lowest percentage in the world. Certainly, sufficient surplus was available to export—3.53 billion dollars worth in the 1950s to 42.6 billion dollars in 1993, an increase of over tenfold, which improved the U.S. balance of trade. American corporations made and sold the fertilizers and pesticides and hybrid seeds and miles of irrigation pipe that helped make this revolution possible.

This revolution was, it seemed, a dream come true, a triumph of American ingenuity and technology. It was the kind of postwar achievement celebrated in those black and white newsreels and "educational" films of the 1950s and 1960s. In the public mind, the story of postwar agriculture was heroic, certainly not tragic.

My organization and I have changed over the years—changed sides, in fact, a rarity in agriculture.

Over the years, as I raised cattle and worked with farmers and ranchers, I slowly came to see, as fellow Oklahoman Paul Harvey would say, "the rest of the story." The three-part indictment, which, if I could, I would nail on the door of every office in the USDA, land-grant university agriculture department, and corporate agribusiness, spells out the truth of the matter.

However, a trial requires evidence, argument, summation. In the rest of this chapter I present the case.

INDICTMENT I

Industrial agriculture stands accused of endangering the essential natural resources of soil, water, and life, thereby jeopardizing the future productivity of agriculture and the inheritance of our children.

Soil

When I was a teenager, I begged my father for three years to plow our fields with a moldboard plow—a plow that dug deep into the ground and turned it completely over. Using a moldboard plow took a big tractor with plenty of horsepower—a symbol of the new industrial agriculture, much desired by a tractor-crazy kid like me. And by turning under the residues—the remaining stalks and leaves from the harvested crop—the moldboard made a "clean" field, much desired by farmers. But my daddy didn't want to do it. The soil would blow away if all the crop residue was turned under, he said. Besides, it took more fuel to plow with a moldboard, and it took longer to plow a field. He eventually gave in and let me plow one part of a field with a moldboard. But he steadfastly stuck to his one-way disk plow, a poor man's plow—which cut up residue and pulverized the soil without turning it completely over, leaving some residue on top.

My father, though I thought he was old-fashioned at the time, was right to resist. Without some crop residues left on the field, the nearly constant winds sweeping down the plains of western Oklahoma threatened to blow our topsoil away. (As it was, the one-way plow did not leave enough residue on the fields to stop the wind damage completely.) Blowing topsoil is like blowing dollar bills; topsoil is where the fertility lies. Without adequate topsoil, which is a mixture of nutrients, organic matter, minerals, and microorganisms, crops cannot be grown.

Although wind erosion was the main threat on the southern Great Plains where I grew up, more erosion is caused by water in the United States. How much erosion is going on? The USDA's Natural Resources Conservation Service (NRCS) has set a "tolerable" rate of soil erosion known as T. This varies by soil type, but is generally between three and five tons per acre per year. The NRCS maintains that erosion rates at this level will not harm soil productivity. By 1992 figures, about 16 percent of the nation's cultivated cropland is losing soil at a rate greater than T due to wind erosion; 21.4 percent greater than T due to sheet and rill erosion, two common forms of erosion caused by water.[1] Although this represents a significant improvement from figures of a decade earlier, it is still a serious problem. (Erosion rates fell slightly more by 1995 and have remained virtually the same since then.) Not accounted for in these figures is erosion from "ephemeral gullies," small channels and swales that come and go, often in highly erodible soils or where soil has been disturbed following harvest of root crops or where there is little or no residue cover. This erosion can add significantly to tallies (in some cases more than doubling them).[2] According to the NRCS, erosion threatens the productive capacity on nearly one of every three cropland acres.[3]

Blowing topsoil is like blowing dollar bills; topsoil is where the fertility lies.

Erosion is not just a problem on croplands—it can affect pastures and rangelands, too. "Accelerated" soil erosion threatens about one-fifth of private rangeland.[4] Besides erosion, other threats to topsoil exist. Overirrigation in western soils can cause them to become saturated with salts, which makes topsoil too salty to grow crops.

This wouldn't matter much if farmers could easily replace topsoil, but they can't. Topsoil forms very slowly in nature, at a rate of one inch each 500 to 1,000 years, depending on geographic location.

Why are farmers endangering the very resource on which they ultimately depend? In short, it is because industrial agriculture requires farmers to exert too much pressure on the soil. Soil, like a rubber band, is resilient but not infinitely so. Pull on it hard for too long and it loses its elasticity. In industrial agriculture today, profit margins are low and risk is high, so maximum production is paramount. These factors have led farmers to adopt farming methods that increase production but may cause erosion or jeopardize the life and health of the soil in some other way.

Some examples: Farmers may attempt to grow crops on "highly erodible" land—land that is too steep or dry or marginal in some way—just to increase production. They may raise too many cattle in too little space, causing the cattle to overgraze and trample the land. Overgrazing leaves the soil bare and open to erosion. Farmers may leave a field bare in the winter rather than plant a "cover crop" to protect the soil from erosion by water and wind, simply because it is expensive to plant, or because they may not realize the importance of cover crops. They may cling to conventional tillage methods, where crop residues are plowed under and fields are left bare and exposed to the elements, which can lead to erosion. However, switching to a different tillage method that would leave residue to protect the soil involves purchasing expensive new machinery and adds to risk and expense. Farmers may plant the same crop over and over, known as monoculture, in order to take advantage of government programs and use specialized machinery. Monoculture of row crops increases soil erosion, which, in turn, depletes the soil of organic matter and nutrients.

Indeed, one of the practices most destructive to soil health is monoculture, growing the same crop in the same field year after year. It is efficient the way that making the same item every day in a factory is efficient. The machinery is ready, the routine is known, and the markets are there. But the soil is not a machine that produces crops. It is organic—alive, not dead. Monoculture depletes the soil of its life and health. Monoculture of row crops in particular increases erosion. Unfortunately, this practice has been heavily supported by government programs that were designed to reduce a farmer's financial risk.[5]

Too often, the shelter belts of trees, the sod waterways, the terraces, and other conservation measures proven to increase moisture and stop erosion in a field are abandoned or destroyed when a farmer deems they are cutting into his profits.

In industrial agriculture's struggle for profit, bigger farms have become the norm, and that, too, affects the soil. Big farms demand big, high-horsepower tractors and heavy farm implements, the kind I was in love with as a kid. As they roll across the fields, their weight compacts the soil, taking out airspaces, and that decreases the capacity of the soil to hold water which, in turn, increases runoff and water erosion. Compaction also negatively affects plant roots. In addition, the financial pressure on farmers to increase their acreage has led them to become responsible for too much land; they don't have the time to implement conservation practices. Neither do they have time when they are forced to work away from the farm to make ends meet, as many farmers must do.

Modern industrial agriculture has, in addition, preached a chemical fix to most soil problems. Fertilizers, it was thought, could make up for loss of fertile topsoil. The long-term effects of such farming practices have not been given enough weight or have been ignored in favor of short-term gain.

Of course, soil erosion is nothing new. During the dust bowl years, when the Great Plains suffered from drought, soil was blown in huge black clouds literally across the country and into the halls of Congress. Myriad soil conservation measures were promoted and adopted in the United States during and after the dust bowl of the 1930s. In recent years, 36.4 million acres have been enrolled in the Conservation Reserve Program and erosion on that land has been drastically reduced.

Although many farmers and government soil conservationists have expended much effort to stop soil erosion, and have met with success, government programs and policies sometimes run at cross purposes. Although some encourage conservation, others have encouraged practices that have led to soil erosion. For example, during the 1970s, farmers were urged to plant as much as possible to supply surging markets overseas. The result was the plowing of land that should not have been plowed. Added to this are the many changes in the names and provisions of the conservation programs themselves. The upshot: too much soil is still being lost.

Too often the shelter belts of trees, the sod waterways, the terraces, and other conservation measures proven to increase moisture and stop erosion in a field are abandoned or destroyed when a farmer deems they are cutting into his profits. This was made clear to me when, on a visit home to Kiowa County for Christmas in1996, I saw a sixty-year-old windbreak of bois d'arc and cedar trees being bulldozed. It made me sick to see it. Trees have been proven to keep wind from drying out and eroding the ground at a distance of twice their height outward. They also provide shelter for wildlife—insects, birds, and other animals that often help the farmer in subtle yet important ways. All for the gain of a few acres of cropland.

It seems indicative of the industrial agriculture mind-set that in neither of the two agricultural dictionaries—one old, one new—on the shelf at the Kerr Center is the word "stewardship" defined. A steward is a careful manager, the person who takes care. Farmers must be stewards of their soil. Too often they are not taking enough care.

Water

It is basic: Sun, soil, and water make life possible. Although good rich soil is becoming more diffucult to find in the fields of America, water, too, is another natural resource equally at risk.

For farmers, the decline in water quality is usually, if not always, an off-farm problem. If the quality of a farm's soil can directly affect a farmer's earnings over the long term, water pollution is less of a direct concern. Although threats to rural wells and drinking water are certainly real, and pollution may impact the bodies of water a farmer may like to fish or swim in, often the effects of industrial farming practices on water quality are felt most downstream.

Agriculture, through runoff from farm fields and feedlots, contributes the most nonpoint water pollution in the United States.[6] The term *nonpoint* refers to sources of pollution that are scattered, with no specific place or point of discharge into a body of water. This makes them more difficult to identify, monitor, and regulate. Nonpoint sources include runoff from golf courses, lawns, roads, and parking lots, in addition to agricultural sources such as fields, pastures, and feedlots. Point sources, in contrast, are specific locations, such as drain pipes.

Just exactly how much of an offender agriculture is remains a subject of some debate, mainly because water quality assessments have been made on only part of the nation's rivers, lakes, and estuaries. How complicated it can be is illustrated by the 1994 EPA assessment of rivers. Only 17 percent of the nation's river miles have been assessed. Of that number, a little over one-third suffered impairment. Agricultural pollution played a part in polluting 60 percent of those.[7] However, taking a narrower view, agriculture played a part in impairing half of the twenty rivers named as the most endangered in 1998.[8]

The city of Tulsa spends $100,000 each year to counter the bad smell and taste in its water supply caused by excess phosphorus.

Eroded soil, fertilizers (both organic and chemical), and pesticides can be pollutants. Soil erosion and water quality are intimately connected. Topsoil washed from fields into bodies of water has a number of negative effects. It can interfere with the life cycles of fish and other aquatic creatures and lead to their decline; it can also fill in ponds, canals, reservoirs, and harbors.

And then there is pollution from organic wastes such as hog and chicken manure. If these organic fertilizers are used too heavily and too often, they can pollute. Phosphorus from chicken litters can attach to soil particles and be washed into lakes and rivers. Excess phosphorus can cause a decline in water quality. This has been identified as a problem in parts of Oklahoma with large numbers of contract chicken farms. The city of Tulsa spends $100,000 each year to counter the bad smell and taste in its water supply caused by excess phosphorus.[9]

Accidental spills and leaks of hog waste from the large open lagoons of hog megafarms can pollute groundwater and surface water.

Pathogens, such as *Pfisteria,* which kill fish and are thought to make people sick, are also associated with these animal wastes in water. Cattle in concentrated numbers allowed free access to streams or ponds can pollute the water with their wastes. In the summer in Oklahoma, this is a common scene—a herd of cattle standing in the shallow water of a pond or creek drinking and cooling off because their pastures have no trees and they are allowed unlimited access to the pond. While standing there, they defecate, and the water soon turns unhealthy, sometimes killing fish.

It has been estimated that as much as 25 percent of the fertilizer spread on farmland each year is lost as runoff. When not all of the nitrogen in fertilizer is utilized by growing plants, runoff containing nitrates can have drastically negative effects on water health. A dramatic example: Each spring and summer a 7,000-square-mile "dead zone" develops in the northern Gulf of Mexico due to this excess nitrogen, which eventually strips the water of oxygen. Fish and shrimp flee the oxygen-poor waters, but bottom-dwellers that cannot leave, such as starfish, simply die. Experts say much of the nitrogen comes from fertilizer washed from fields in the Midwest which then flows down the Mississippi River.[10] According to the NRCS, the "average annual nitrate-nitrogen concentration in the river has doubled since 1950."[11]

Nitrates can reduce the oxygen-carrying ability of blood, especially, but not only, in fetuses, babies, and young children. According to the U.S. Geological Survey, wells in agricultural areas are most at risk. . . .

Since 1979, the agricultural sector has accounted for about 80 percent of all pesticide use each year. Pesticides, drifting into watersheds or running off fields, can cause large-scale fish kills, such as a recent incident on Big Nance Creek, a tributary of the Tennessee River in Alabama. State agency personnel estimated that 240,000 fish were killed by a build-up of pesticides in the water.[12] Trout fishermen take note: Rainbow trout are reportedly the fish most sensitive to pesticides.

These same nitrates and pesticides that can wipe out aquatic life can affect human health via drinking water, if found in high enough concentrations. Nitrates can reduce the oxygen-carrying ability of blood, especially, but not only, in fetuses, babies, and young children. According to the U.S. Geological Survey, wells in agricultural areas are most at risk, with over 20 percent of shallow drinking wells in agricultural areas exceeding the safe limit, compared to 1 percent of public wells. Nitrate concentrations are highest in areas dominated by row crops and having well-drained soils.[13]

Industrial agriculture has had an impact not only on water quality, but water quantity. Industrial agriculture's push for higher production has encouraged the mining of water from underground sources—withdrawing it at a high rate without regard for the long term.

A case in point: About 100 miles northwest of where I grew up in Cold Springs, the Ogallala Aquifer slips silently into Oklahoma. It is a three-million-year-old vast underground river, although some describe it as more like a gigantic wet sponge than a river. It underlies parts of eight states in the Great Plains, from Texas in the south to South Dakota in the north, and eastern Kansas to Wyoming—hence its other name, the High Plains Aquifer. About 30 percent of the groundwater used for irrigation in the United States is pumped from the Ogallala, to irrigate about 14 million acres (in 1990).

Using the Ogallala's water has transformed agriculture in the region since World War II. Farmers went from raising dryland wheat and grain sorghum and being at the mercy of all-too-frequent drought (part of this area was the old dust bowl) to irrigation agriculture—raising corn, sugar beets, alfalfa, and cotton. Raising corn especially helped supply feed for the cattle being fattened at the many commercial cattle feedlots and packinghouses that sprang up on the southern Great Plains. The idea of irrigation as insurance during times of drought gave way to regular heavy use.[14]

The problem is that the Ogallala underlies an area of the country that is dry, and water is being pumped out faster than it is going in. Water levels have dropped significantly—in some areas as much as 100 feet. Wells have had to be sunk ever deeper; the cost has become prohibitive for some farmers. With more efficient irrigation practices, producers are able to use less water, and the rate of decline of the water level has decreased. But when drought strikes, as it did in the mid-1990s, water levels begin to drop quickly again.[15] Though no one can predict the exact day it will happen, the Ogallala will be lost as a resource for future generations if water keeps being withdrawn at such excessive rates.

Life

Sun, soil, and water make plants grow. If soil and water are endangered, what about the sun? Well, with five billion years down, five billion to go in its life cycle, it's not at risk. However, the plants that capture the sun's energy in photosynthesis are a different story. They are at risk, as are the animals that eat them, one step up in the food chain.

About half of the land in the United States is privately owned crop-, pasture-, and rangeland. It is fair, then, to say that farmers and ranchers have a significant impact on the amount and diversity of life-forms found in the country. Agriculture has tamed wild America; wild plants and creatures declined as wetlands were drained, forests were felled, and prairies were plowed for agriculture. Industrial agriculture, with its emphasis on increasing production, often demands that every square inch of a farm or ranch be "put to good use," leaving little room for wildlife habitat. Pesticides, too, can and have harmed wildlife. That wild plants and animals have an ecological as well as aesthetic value on the farm, as well as value to society at large, is not part of the industrial agriculture equation.

Although the cause of beleaguered wildlife has been taken up by conservationists, the plight of endangered domesticated plants and animals is not as well publicized. Few know just how genetically uniform our crops and livestock have become.

Thousands of old varieties—locally adapted and containing a wide array of valuable genetic characteristics that may one day prove useful—have been abandoned in favor of higher-yielding hybrids, and, in many cases, lost forever.

Corn is a case in point. Hybrid corns began to be adopted in the 1920s; by 1960, hybrid corns were planted on 96 percent of corn acreage. In the 1960s, scientists introduced varieties of wheat and rice that were much higher yielding than earlier varieties. This was the green revolution, and its leading scientist, Norman Borlaug, received the Nobel Peace Prize in 1970 for his work in developing high-yielding wheat. In the past forty years, hybrid grains have made modern agriculture spectacularly productive. Though the casual observer is not aware of it, there is a reason why the corn plants in a field are all the same height, form ears at the same time, and have such high yields: they have been bred to respond to chemical fertilizers and have uniform growth habits adapted to farm machinery.

The adoption of these new wonder varieties has not been completely positive, however. Thousands of old varieties—locally adapted and containing a wide array of valuable genetic characteristics that may one day prove useful—have been abandoned in favor of higher-yielding hybrids, and, in many cases, lost forever.[16] According to the United Nations Food and Agricultural Organization (FAO), about 75 percent of the genetic diversity of agricultural crops has been lost![17] With the recent introduction of genetically modified grain seed that a farmer cannot legally replant, the genetic pool seems destined to narrow further.

Genetic diversity is necessary insurance for the future—against changes in climate, such as possible global warming, evolving pests and diseases, and changes in the availability of energy,[18] along with changes in fashion, markets, and the structure of agriculture. Genetic diversity is essential if the increasing population of the world is to be fed.

Uniformity in crops and livestock threatens our food security.

History has shown that diversity is good, and that uniformity in crops makes them vulnerable to disease. The Irish potato famine of the 1840s, which caused massive starvation and displacement was caused by such genetic uniformity. The potatoes grown in Ireland were the descendants of just a few varieties of potatoes brought back to Europe from the New World by early explorers. Unfortunately, these varieties were susceptible to *Phytophtora infestans,* a blight that wiped out most of the crop for a few years. In more recent times, 15 percent of the corn crop in 1970 was lost because of a blight. The common hybrids were all descended from a blight-susceptible variety.

Many breeds of farm animals, too, are declining in number. According to the American Livestock Breeds Conservancy, nearly eighty livestock breeds in North America are in decline or in danger of extinction. The situation of poultry varieties is similarly dire.

This doesn't mean that these breeds or varieties are inferior in all characteristics to the ones currently popular. These days, both individuals within a breed and breeds within a species are being selected toward the highest-producing type.[19] This tendency is obvious in the Confined Animal Feeding Operations, or CAFOs, that produce almost all of the chicken and turkey (and a quickly increasing percentage of the pork) in the United States. The most efficient, uniform type of bird is raised in tight quarters in these highly controlled environments. Uniformly sized animals are important to industrial agriculture because they make assemblyline slaughterhouses more efficient.

Beef cattle, sheep, and swine are all selected for a standard carcass profile, rapid weight gain, and ability to utilize high-concentrate feed. Breeds that deviate from this production ideal are discarded. This is made easier with the use of artificial insemination, embryo transfer, and other high-tech reproductive techniques. The result: fewer animals are used in breeding. The genetic base is weakened when animals that are climate adapted or show strong maternal instincts—traits that are often not relevant to highly industrialized production of animals—become extinct.[20]

Uniformity in crops and livestock threatens our food security. History has shown that preserving many varieties of plants and breeds of animals—in other words, preserving diversity—can save the day. Blight-resistant varieties of potatoes that were still being grown in Mexico and the Andean countries allowed the potato to throw off the Irish blight and become a major food crop today.[21] A largely ignored Turkish wheat was found to have resistance to a disease that threatened the wheat crop in the northwestern United States in the 1960s.[22] Animal breeds also differ in their resistance to diseases and parasites, as well as in many other traits that may be important in the future, both commercially and environmentally.

Our Children's Inheritance

In a world where the U.S. population is expected to reach 335 million by 2025 and global population is expected to reach nearly eight billion by 2020 (about 25 percent more than today)[23] the danger to our food supply from soil degradation, water depletion, and loss of plant and animal varieties is real. All over the world, the natural resource base on which agriculture depends is threatened.[24]

The bottom line: Agriculturists must protect, even enhance, our natural resources. We can't afford to abandon, exhaust, or contaminate our storehouse of finite natural resources.

INDICTMENT II

Industrial agriculture stands accused of hooking farmers on fossil fuels and the fertilizer and pesticides made from them, while downplaying the consequences of overusing such products.

After I graduated from high school in 1965, I tried to follow my heart and be involved in agriculture, but I also wanted to please my mother, who wanted a different life for me. So I majored in agriculture in college, intending not to farm full time, but to teach vocational agriculture at the high-school level. But after getting my master's degree in agricultural economics in 1972, I was hired by the agricultural division of the Kerr Foundation in Poteau, Oklahoma, some 300 miles east of where I had grown up. My job was to visit farms in the southeastern part of Oklahoma as part of a consultation team. I helped farmers and ranchers make decisions about farm management and financial strategies. So I remained much more in touch with real farming than I could have done teaching agriculture.

As Kerr specialists, we dispensed what we thought was sound advice to the farmers of our county, Le Flore, and surrounding counties. It was conventional advice, generally agreed upon by county extension agents and university agriculture professors. The soils of Le Flore County, depleted from deforestation and poor agricultural practices, needed some enriching, so our specialists prescribed judicious fertilizer applications. Against weeds, herbicides were the ticket; and against other pests, the other 'cides, fungicides and insecticides. I advised farmers and ranchers how to borrow the money to pay for it all.

Chemicals had not always played such a big role in growing food. During World War II, the petrochemical industry, an offshoot of oil refining, expanded. After the war, the industry changed munitions plants into plants to manufacture ammonium nitrate fertilizer. Other agricultural chemicals, such as pesticides, were also manufactured cheaply—in some cases created from leftover nerve and mustard gases. Farming entered what has been termed the chemical age.

Because these chemicals were cheap, widely available, and increased production, agriculture became utterly dependent on them. Take fertilizer. Ammonium nitrate, a key component of fertilizer, is made from ammonia, which is made from either natural gas or naptha, which itself is made out of oil. It supplies an essential plant nutrient, nitrogen, and it is vital to industrial agriculture: American farmers used 2.7 million nutrient tons of nitrogen a year in 1960; by 1995 they were using 11.7 million tons.[25] Use continues to be heavy. I submit that any given farm acre in America is as tied to oil and natural gas as any urban acre, and the role of these nonrenewable fuels in crop production has become as important as those of soil, water, and sun.

In 1995, 97 percent of U.S. corn acres and 87 percent of wheat were fertilized with nitrogen.[26] These two major crops indicate the extent of fertilizer use. With the advent of cheap, easy-to-use fertilizer, old methods of enhancing soil fertility, such as planting a legume (bean) crop in rotation with other crops, fell out of favor. (Legumes draw nitrogen from the air and add it to the soil.)

Other petrochemicals such as herbicides, which kill weeds, are almost as liberally used; since World War II farmers have become ever more reliant upon them. In 1988, 96 percent of corn acres were sprayed with herbicides (up from 79 percent in 1971), as well as 96 percent of soybean acres (up from 68 percent). How much herbicide is that? In 1989, the cornfields of America were projected to take 219 million pounds of herbicides. And insecticides, too, were relied upon, though not as heavily. They were used on about one-third of the corn acres.[27]

To sum it up, the United States is the largest user of pesticides in the world, and agriculture uses 75 percent of the pesticides in the United States.[28]

In industrial agriculture, chemicals are an intrinsic part of the system, and their use has been advised by almost everyone. Only a few people (the ones on the plaintiffs' side) have thought very much about the consequences of their use.

Because these chemicals were cheap, widely available, and increased production, agriculture became utterly dependent on them.

This advice is expensive. Tallying it up, fertilizer accounts for about 30 percent of U.S. farm energy use.[29] On the farm, there are myriad uses for fossil fuels (gasoline, diesel, propane) or products made from them. The most important direct use is in powering tractors and other farm machinery. Fossil fuels also run irrigation pumps, the trucks that carry crops to market, and the trucks that carry seed, fertilizer, and other chemicals to the farm. Propane or natural gas are used to dry some crops and used to heat the CAFOs, such as the pig and broiler houses that produce the meat for urban tables.

For every unit of food energy eaten in the United States, nearly ten units of energy are spent producing it, processing it, and shipping it to our tables.[30] Much of that energy is used in the extensive processing, packaging, and distribution system.

And then there are the less direct but very important uses of fossil fuels: The energy used to manufacture all the chemicals, farm implements, tractors, and processing equipment.

Where Is It All Headed?

As we say in Oklahoma, where's it all headed? Unfortunately, it looks like it's headed downhill, putting the nation's most important industry, agriculture, at risk. Oil and gas (and coal), the fossil fuels, were formed very slowly deep underground millions of years ago; they cannot be renewed, and will eventually run out. Crude oil has never been manufactured in a lab. Analysts argue over when that might be: oil in thirty to fifty years, natural gas in sixty to 120 years. How fast the reserves are pumped dry depends in part on how much newly developing countries use and in part on how well we are able to exploit reserves, which are often in environmentally sensitive areas such as the Arctic Wildlife Refuge in Alaska. Some analysts predict that, within the next decade, supply will not be able to keep up with demand.

Before the tap turns off, though, we will become more and more dependent on foreign supplies—largely from the Middle East, a notoriously unstable part of the world. Net imports of petroleum are expected to reach 65 percent of domestic petroleum consumption by 2020. Imports of natural gas are also expected to grow, as is consumption (by as much as 50 percent by 2020).[31]

With demand high (and increasing) and supplies limited, the price will inevitably go up. For farmers dependent upon these scarce resources, such price increases will be bankrupting. The farmer will be caught in a bind—the cost of inputs needed to make a crop steadily rising, while the prices received for the crop are stagnating.

Some predict that in twenty-five years, farmers may find energy to be unaffordable.[32] If so, how will farmers survive? The process of getting hooked on fossil fuels is sadly similar to drug addiction: The farmer becomes more and more dependent, spending more and more money and unable to farm without chemicals, while in denial over the havoc that this habit has wreaked. For instance, using large amounts of chemical fertilizer, as is the norm, sets up a vicious cycle. By adding fertilizer, you can work the soil more intensely without decreasing yields. The fertilizer increases the yield, but also masks the decline of organic matter, which is important to a healthy soil.

The heavy use of herbicides and insecticides brings its own vicious cycle. Overuse of insecticides has wiped out beneficial insects along with harmful ones, while at the same time leaving pests resistant to the chemicals. Currently, more than 900 insect pests, weed species, and plant diseases are resistant to one or more pesticides.[33]

Of course, the act of applying pesticides can threaten the health of farmers. Studies have shown increased rates of some cancers among farmers. Although cancer may take years to develop, farmers and farm workers also can suffer sudden, acute poisoning—with as many as 20,000 deaths worldwide from pesticide poisoning each year.[34] Farming as an occupation can be hazardous to one's health.[35]

And the hazards, although perhaps most acute on the farm, extend far beyond it. Agricultural chemicals have had and continue to have numerous adverse effects on wildlife, and, as noted earlier, on water. Also, consumers worry about the long-term effects of small amounts of pesticide residue on food (which hasn't been studied).

What kind of agriculture would we have today if we had asked different questions and researched different solutions?

As we also say in Oklahoma, how'd it get this far? When it came to recognizing these dangers, where were the legions of agricultural scientists and educators that populate our land-grant universities? They seemed compelled, as I have heard many times, to "go where the funding is." And some research funds come from agribusinesses, which naturally are mostly interested in having farmers buy their products. In 1992, the private sector provided $143 million to state agricultural experiment stations and their cooperating institutions.[36] And USDA research money has also largely served the status quo. The result is that basic research into alternatives to these chemicals has often been ignored. Not that researchers were ill intentioned; they believed (and still do believe) that industrial agriculture is the only way to go. And the tide certainly runs that way: for research in 1992, the private sector outspent the state and federal governments by 1.2 billion dollars. Research into agricultural chemicals accounted for 37 percent of the total.[37]

Although a new attitude, new research, and new solutions are emerging, the fact remains that only a fraction of the billions of research dollars spent since World War II has gone to projects looking into alternatives to petrochemicals and the industrial approach. Currently, only about 15 million dollars per year goes to fund the USDA's SARE program, which researches innovative alternatives. The bias continues, and I maintain it is a case of criminal negligence.

I wonder what kind of agriculture would we have today if we had asked different questions and researched different solutions? Perhaps we would have an agriculture that would have been productive, yet kinder to the earth, its wild creatures, and the small farmers who are, as Jefferson said, the most precious part of the state.

INDICTMENT III

Industrial agriculture stands accused of desolating rural America by bankrupting farmers and ignoring the well-being of rural communities, thus leaving them open to exploitation.

Not long after I moved to Poteau to become a consultant, I became a part-time rancher. I borrowed money to buy five acres and eight cows. This was my first farm, such as it was.

The fact that I raised cattle (and my farming background) lent me credibility with area farmers as I advised them on how to get and manage the gigantic amounts of money it took to farm, guiding them as best I could through the maze of USDA farm programs.

It takes a lot of money to farm because farms today have become very big. The average farm size is now about 470 acres, up from 175 acres in 1940 (see Figure 1.1). And this figure is misleading because most of the food is grown on farms that are much larger than this average. In 1993, almost half of the gross farm sales came from farms where the mean number of acres was about 3,000.[38]

Family labor is not adequate for these larger operations and has largely been replaced by expensive farm machinery, such as the tractor. It may be hard for younger readers to understand that at the end of

FIGURE 1.1. Average Number of Acres per Farm in the United States, 1940-1997

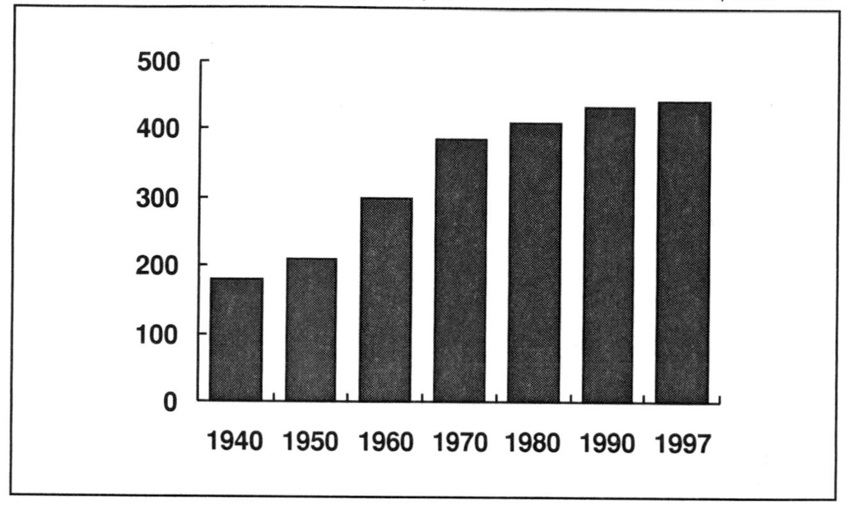

Source: Data from *History of American Agriculture 1776-1990,* USDA, Economic Research Service, Washington, DC, POST 11, and USDA.

World War II only 30 percent of farmers used tractors—the rest still relied on horses, mules, and human labor.[39] That quickly changed in the 1950s. Tractors allow a single farmer to plant and harvest more acres but tractors are a big expense, both to purchase and to maintain. Farmers must also have implements to use on the tractor, such as plows, discs, planters, and drills, to name just a few. And then there is additional specialized harvesting equipment, such as grain combines.

Other big costs in farming today are what have become essential inputs, such as fertilizer and pesticides. The equation of big acreage plus big machines plus big chemicals has, since World War II, added up to big production. Agricultural research at universities boosted the push toward greater production by conducting tests of new higher-yielding varieties and new chemicals that were designed to do just that—boost production.

The problem is that big production depresses commodity prices. It is the old law of supply and demand. In post–World War II agriculture, farmers got caught in a vicious cycle. Producers felt they must get bigger and produce more to make a profit, but doing that kept prices low. By and large, USDA farm programs have been unable to slow this trend of overproduction. And the promise of expanding export markets that will buy our overproduction has yet to happen on a regular basis.

The upshot: the latter half of the twentieth century has been a time of consolidation, what might be called the era of the vanishing farm. In 1950, there were 5.3 million farms; by 1960, down to 3.7 million; by 1980, 2.4 million; and in 1997, about 2 million. In short: 62 percent of American farms have disappeared since I was a toddler. When farms disappear, so do the families that farm them—in 1940, farmers made up 18 percent of the labor force; in 1995, they were down to 2.7 percent. If we were a bird species, some might deem us eligible for the endangered species list (see Figure 1.2).

Even the figure of two million farms is misleading because one need only have $1,000 in farm sales to be classified as a farm. Only 8 percent of all farm operator households receive income from farming at or above the average income for U.S. households.[40]

As is true for families all over rural America, farming cannot begin to support my wife and me and our two children. In our case, it's been "town

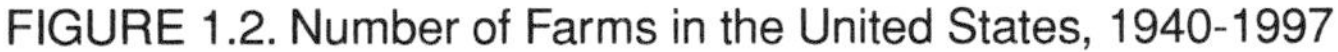
FIGURE 1.2. Number of Farms in the United States, 1940-1997

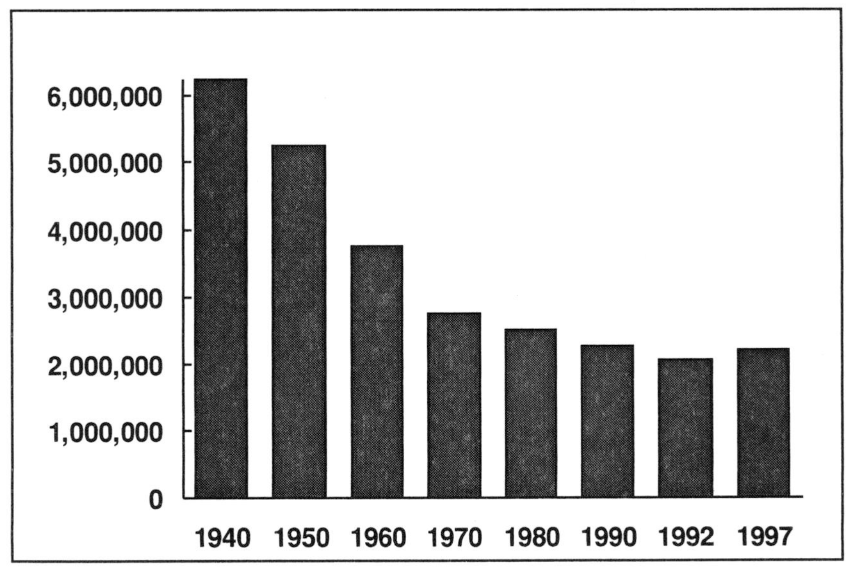

Source: Data from *History of American Agriculture 1776-1990,* USDA, Economic Research Service, Washington, DC, POST 11, and USDA.

jobs" (my wife is a teacher) that have financed our land and home. Over the years, we have been able to buy 240 acres and we now own forty cows. We have been able to make enough from the cattle operation to build barns and corrals and buy farm equipment, though in some years it is tough to make a profit.

When farms disappear so do the families that farm them—in 1940, farmers made up 18 percent of the labor force; in 1995, they were down to 2.7 percent. If we were a bird species, some might deem us eligible for the endangered species list.

Where was the government through all these changes? Acknowledging the difficulties small farmers were having surviving during the Great Depression, the government first stepped in during the 1930s to support farm prices and has been deeply involved in agriculture ever since. However, the results for the small farmer have been disastrous in the long run. At a steep cost to the nonfarming taxpayer, farm subsidies "have contributed to uneven income distributions" among farmers, favoring large landowners. In fact, in 1992, 68 percent of the U.S. government farm payments went to the wealthiest 19 percent of agricultural producers.[41] African-American farmers, often smallholders, were hit even harder as a result of discrimination, for which the USDA is now paying restitution.

Boom and Bust

I did not question the wisdom of "bigger is better" as a young agricultural economist, fresh out of graduate school, feeling my oats. The 1970s seemed to me to be the American farmer's glory days. It was a time of expansion, encouraged by the ease with which a farmer could borrow money from the Farmer's Home Administration. The inflation in the country as a whole drove up the price of land; if you owned land, you could borrow against it to buy more land and the bigger tractors and farm implements with which to farm it. The pattern that my family had followed—farming for years as sharecroppers in order to save enough money to buy a place—had become as rare as that no-frills little tractor we used, the two-cylinder Poppin' Johnny. *Saturday Night Fever* hit urban America, and big farm and big tractor fever gripped rural America.

What was it like then? In the early 1970s, then Secretary of Agriculture Earl Butz encouraged farmers to expand, to plant fence row to fence row and even in the ditches in order to supply the export market to such new customers as China and the Soviet Union. It seemed like a business opportunity that was too good to turn down. Everything was increasing in value—a farmer could buy a new tractor, use it for three or four years, and sell it for more than he had paid for it. And land! The great thing about a land boom when prices are increasing every year is that it is a form of self-financing. As the value of land increases, so does farmers' buying power. Since they are worth more, they can borrow more. This leveraged buying became the rule. I remember recommending it, and farm magazines encouraged it. I could walk in with almost any farmer to a Farmer's Home Administration office and help him get a forty-year mortgage on a piece of land, as well as a new house built on it at 5 percent interest. If he had any further financial difficulty, we could rework his farm plan and get his loan refinanced at the same rate. There were income tax deductions for interest; investment credits encouraging the purchase of expensive machinery; everything was in place to make agriculture the place to be in the 1970s.

There were, I saw, losers in the midst of plenty—small farmers who wanted to buy small pieces of land were passed over in favor of the big boys who needed megaloans. The economies of scale ruled the day.

With every boom, however, there comes a bust. In the latter part of the decade, the fantasy of farming profitably-ever-after ended. Again, the government was heavily involved. The Federal Reserve decided to fight runaway inflation with a tight money policy. Interest rates rose, making it difficult for farmers to pay back the loans they depended on, and exports slowed. A grain embargo was instituted against the Soviet Union. Meanwhile, farmers, with their big tractors and economies of scale, were producing record amounts of grain. Prices dove. The prices of everything—tractors, fertilizer, all the paraphernalia of big farming—had been going up. In a period of about five to seven years, nearly everything was undone. Farmers faced debts they could not pay off.

What happened in the state of Iowa is a good illustration of events. In the 1980s, Iowa's farm population dropped 34 percent, with nearly 35,000 people leaving farming. Similar declines were found in other Midwestern farm states, such as Illinois, Missouri, and Minnesota.[42]

Only 8 percent of all farm operator households receive income from farming at or above the average income for U.S. households.

The farm crisis of the 1980s, dramatic as it was, was just the latest episode of agricultural disintegration and rural depression. Everyone who has grown up on a farm has seen it. In the community where I grew up, within a four-mile radius of our farm, there were sixteen families in 1955. By 1965, only two remained.

In reality, the farm crisis is ongoing. In our headlong rush to industrialize agriculture, we did not foresee some of the results: "social disorganization, shrinking rural economic bases, declining rural communities and institutions, and the specter of a permanent underclass in the city," as University of Missouri agricultural economist John Ikerd has written.[43] With urban America's general disdain for rural life (we are hayseeds or hicks, living in Podunk towns, who would never stay down on the farm after seeing Paree), an important truth has been missed: having viable rural communities is very important to the health of the nation as a whole.

Unfortunately, in Oklahoma, as in many farm states, rural communities are struggling. In the county I grew up in, Kiowa, and the county in which I now live, Le Flore, poverty rates are well above state levels, and per capita incomes lag behind the state average.[44] While our national attention has been focused on the urban poor, the rural poor have been overlooked. By the late twentieth century, 25 percent of all rural children lived in poverty, and, overall, the poverty rate of rural America exceeded the poverty rate of the cities.[45]

Small and even medium-sized farms, owned by families and worked largely by family labor, do not have the financial resources to survive many bad years in a row, whether their losses are due to low prices or to natural disaster. Farm statistics tell the tale: 73 percent of the farms in the country receive net farm incomes of around $1,000 per year. They account for only 10 percent of gross sales, and are appropriately classified as "noncommercial."

Examined from the other end, 27 percent of farms in this country account for 90 percent of gross sales. However, less than 1 percent of all farms—the superlarge with $1 million or more in sales—accounted for just over 25 percent of all gross sales.[46]

What are the effects of this dichotomy? "As farm size and absentee ownership increase, social conditions in the local community deteriorate," says one analysis of the effects of "get big or get out." Farm communities surrounded by farms larger than what can be operated by a family unit reportedly have "bi-modal incomes, with a few wealthy elites, and a majority of poor laborers, and virtually no middle class. The absence of a middle class at the community level has a serious negative effect on both the quality and quantity of social and commercial services, public education local government, etc."[47]

Contracts and Concentration

Farmers with smallholdings are faced with few opportunities for making a living from agriculture. In some areas of the country, including my area of Oklahoma, many small farmers are signing contracts to produce poultry and hogs for large corporations. Although these contracts allow farmers to stay on the farm and have other positives, they have many negatives. Often the contracts are short term, their terms cannot be negotiated, and they require farmers to take out huge loans with unreasonable financial risks for the sake of minimal financial gain. Those wanting to raise chickens cannot do so independently; they are instead tied to a big company. Although they are called independent contractors, they are, in effect, much more like hired hands, and, even worse, hired hands who are deeply in debt. Farmers in these situations have lost much of their decision-making power.

While our national attention has been focused on the urban poor, the rural poor have been overlooked.

These contract chickens and hogs are raised in CAFOs, captive in large buildings where they are raised in crowded conditions, fed expensive high-energy grain, and given regular doses of antibiotics to keep diseases at bay, a must in such close quarters. The huge volumes of waste from such operations can easily turn into a pollution problem.

How times have changed. We raised a few pigs and cows on our farm when I was growing up, and chickens too. These enterprises were different then. Our chickens ranged free, and our hogs had plenty of space. Raising chickens gave the farm family small amounts of cash for current expenses and provided light work for children. In our family, selling eggs was part of our weekly trip to the town laundromat. We delivered eggs while the clothes were being washed, making four or five dollars each time. It seemed at the time a significant amount of money, and indeed it helped the cash flow between our two big harvests of cotton in the fall and wheat in the spring. As for hens who had stopped laying, we sold them at the local farmers' cooperative. I know of one family whose son's college education was partially financed through selling eggs.

Little did I know at the time, but these were the final days of these little enterprises. Today, an open wholesale market for chickens does not exist. The independent farmer can produce chickens or eggs, but can only sell them directly to consumers at farmers' markets or at farm stands. Otherwise, there is no open market for their sale. I am sure that if I had a flock of chickens and needed to sell them that I would not be able to interest a large chicken-producing company in buying them.

Why not? Such companies are consummate practitioners of industrial agriculture, complete with vertical integration of production, processing, and distribution. Although it is true that nonfamily corporations own only .4 percent of farms, this statistic does not reflect how much control corporations exercise over agriculture. Production of vegetables for processing, of potatoes, of sugar beets, of seed crops, of eggs, of broilers, and of turkeys is largely done under contract. The marketing of other commodities, in particular fluid milk and citrus fruits, is done by contract.[48]

This might not be so bad if there were many companies vying for what the farmer has to offer. The opposite, though, is occurring. Control of agricultural marketing and production is being concentrated. Just a few examples: Four beef packers control 72 percent of the market. In 1989, twenty feedlots marketed more than 50 percent of fed beef. Four companies control almost half of the broilers; five companies, 40 percent of the turkeys; five companies, over 60 percent of flour milling. And in these cases and others, the percentage controlled is steadily going up.[49] Just as alarming are predictions that traditional seed companies will soon disappear, and in their place will be a few "consortiums" that will control every step in the "food chain," from seed to final product on the grocery shelf.[50]

As the editors of *Farm Aid News* have written: "Shrinking competition in the food industry has enabled agribusinesses to pay farmers below their costs of production for raw food products while increasing the prices consumers pay at the retail level. Meanwhile, agribusiness corporations that now control the transport, processing, marketing, and retailing of our food are enjoying record profits. America's farmers and consumers are being short-changed."[51]

Communities must accept the fact that the wealth created by such "farms" largely goes out of the community to corporate headquarters, by and large out of state.

Besides the increase in production and marketing contracts in agriculture, some companies, notably those involved in raising hogs, are doing away with farmers altogether and doing it all themselves in huge factory farms where tens of thousands of animals are produced. The problems of odor and waste disposal for these operations are often formidable. Usually located in remote areas, these farms and the slaughter plants associated with them pay low wages and attract immigrant workers. Residents by and large do not take these jobs but are expected to foot the bill for the increased costs associated with such operations, including paying for more schools and special instruction for non-English-speaking children, as well as increased infrastructure costs. Communities must accept the fact that the wealth created by such "farms" largely goes out of the community to corporate headquarters, by and large out of state.

And sometimes residents realize too late that they are stuck with the bill from the pollution these enterprises cause. Lake Wister, the water supply for Poteau and most of Le Flore county, is in critical condition, one of several lakes in Oklahoma in trouble due to excess phosphorus. The phosphorus comes from the chicken manure produced by contract chicken houses (and chicken processing plants upstream). And this is not just in Oklahoma; such pollution is degrading rivers and lakes throughout the South.

Such is the state of farming today. Desperate farmers and rural communities, devastated by farm failures, are embracing these enterprises. I call it desperation economics.

But it goes beyond economics. Running agriculture this way has a moral cost, too. Not only are farmers being exploited, but the animals are as well. The short lives of these animals have no dignity; they are confined in tiny spaces and treated more like inanimate objects than living creatures. Those viewing the transportation of CAFO chickens to market—jammed one on top of the other with no room to move—must ask themselves if cheap chicken is worth this. At some point, our society has to decide whether it wants to apply the industrial model to every enterprise—whether, in fact, we want animals to be "manufactured" like VCRs and farmers to be like factory workers.

With the disappearance of farming as an available independent livelihood, more is lost than just a job, as important as jobs may be. For time immemorial, an unwritten contract has existed between the farmer and the next generation—that they shall receive the legacy of the land and the knowledge of how to work it and care for it. The loss of the family farm is more than just a sociological loss—it's a spiritual loss that reaches beyond the farm and affects society as a whole.

Once, many urban Americans had at least some tie to the land—a parent, grandparent, or other relative farming. Farms used to ring cities, giving city folks a chance to visit them. Now, these agricultural lands are being taken by urban sprawl—subdivisions and shopping centers—but a longing remains for a connection to farm and farmer, a connection to the good earth. Children, making mud pies and digging for treasure, are drawn to dirt practically from day one. People often don't outgrow the fascination with soil and growing things. Farmers' markets are booming, in part because people desire a closer connection to the source of their food. One big city restaurant found that even planting a few living, growing corn plants improved business: "I had people pick their own," said the chef. "They loved it. I guess they had never picked corn before."[52]

The loss of the family farm is more than just a sociological loss—it's a spiritual loss that reaches beyond the farm and affects society as a whole.

What happens when urbanites do not get this contact with farm and farmer? Early in this century, the great American horticulturist and educator, Liberty Hyde Bailey, remarking on the steadily declining ratio of farmers in American society, pointed out that fewer and fewer people in society would be brought in touch with the earth in any real and meaningful way. He also suggested that the shift of farming from its "agri-cultural" roots to an "agri-business" mind-set, risked the loss of this basic human endeavor as a means of "spiritual contact" with creator and creation. This lack of contact with the natural world, he suggested, fosters a basic attitude of contempt toward creation.[53]

THE SOLUTION

> *These crimes show a reckless disregard for the life and health of farmers, rural communities, and the natural world, jeopardizing our ability to feed an ever-growing population. As a result, the food security of our nation, and our world, is threatened.*

These are the main arguments in my case against industrial agriculture. I believe I have set forth the outline of a strong case: that what we have today is an unhealthy agriculture that cannot endure. In the rest of the book I elaborate on the argument and present a path to an agriculture that can be sustained in the twenty-first century and beyond.

What form will this "sustainable" agriculture take? It will have to be an agricultural system that that not only keeps farmers on the farm, makes them a profit, and encourages a free market, but also preserves natural resources and the health of the natural environment. It will be an agriculture that creates healthy rural communities and helps renew society's spiritual contact with the earth. To make this new kind of agriculture will require a new kind of green revolution.

Industrial agriculture has failed and will continue to fail. Change is desperately needed. If I ever doubt the certainty of this, I need only drive back to Kiowa County. Going back home, I see stores boarded up and other signs of decay. Most of my friends—those boys that, like me, were in love with tractors left long ago, like I did, our folks pushing us to pursue "a better life." Our fathers and mothers could see that we were nearing the end of a way of life that in some ways was impoverished and difficult, and in other ways was rich and meaningful. As it turns out, the farming life we knew was not transmuted into something less difficult yet equally laden with value; it simply disappeared for most of us.

Sometimes I wonder what I might have become if I had been one of the few who had stayed. Would I have taken the expected road and perhaps gone through bankruptcy in the 1980s? As it was, after some years of doing what was conventionally thought wise, I picked the path less traveled and it truly has made a difference.

Chapter 2

The Remedy: Sustainable Agriculture

> Sustainability of an agriculture that is environmentally benign in relation to world resources, population, and environment is a serious issue—perhaps, along with population, the central issue for the human race.
>
> John Pesek[1]

We need revolutionary thinking to rescue this nation's agriculture from its own success. So I said in an April 1987 letter to Congressman Jamie Whitten a few days before I testified to his subcommittee on agricultural appropriations.

Revolutionary thinking? I never thought I would write that phrase, especially to a representative of the United States government. I was, after all, the same person who once had practiced giving anticommunist speeches beside the barn for high school Future Farmers of America competitions, and I still consider myself a conservative. In fact, I know that although I have generally been willing to entertain new ideas, I have been slow to adopt them.

But, over time, I became convinced that the statistical success of American agriculture was misleading. Facts such as one American farmer provides food and fiber for sixty, or eighty, or, currently, 129 people, are often trotted out to dazzle the listener and demonstrate the might of American agriculture. But these facts do not tell the whole story. As I wrote to Whitten, such a statistic does not even hint at "the massive financial distress in agriculture today, the number of farm foreclosures, and the number of suicides and stress-related illnesses among our farm people."

If there was one thing I learned in my fifteen years as an agricultural economist in Oklahoma, it was that things in agriculture were not as they appeared. Despite the sunny productivity and balance of trade figures, storms were often brewing on the horizon. Another thing I learned in Oklahoma was that such storms can come up mighty fast and often hidden in their hearts are tornados. The 1980s' farm crisis had been such a tornado and many farmers had been swept up in it—and were lost.

I was going to Washington to propose funding for a solid shelter against future storms: a program that would research resource-efficient or low-input sustainable agriculture (LISA), soon to be known by its acronym. The LISA program would establish four regional centers that would give grants for research and extension projects. These projects would look at ways that the farmer could reduce the amount spent on fertilizers, pesticides, feed, fuel, labor, machinery, purchase of capital, and purchase of information. The reasoning behind this was simple: Farmers have limited control over the price they get for crops at market, but they do have control over the amount they spend to make a crop. By decreasing the use of chemicals and other expensive items purchased off the farm, supporters of LISA believed that the farmer's bank account would be less strained.

These centers were especially needed because what we were proposing was not simply a modification to the conventional approach by cutting out a spray here or there. We were proposing the testing of whole new ways to grow crops and raise livestock, new "systems" that simply didn't need a lot of purchased inputs—chemicals, fertilizers, fuel—to be productive. Such systems would help preserve the health of natural resources and the environment. By providing farmers with information they desperately needed on innovative practices, on designing new production systems, and on the economics of the changes, the regional centers would help make farms sustainable. And it wouldn't cost a lot by USDA standards. Most of the total 30 million dollars would provide grants to farmers and university researchers.

In recommending this program, we asserted that the USDA had not been keeping up with current research and needed to be more involved in exploring low-input ideas. This kind of research was ongoing at private research centers and at some state agricultural stations. Some farmers who had tried new approaches had been successful. The USDA itself had studied organic farmers—those who had abandoned synthetic fertilizers and pesticides—and had found "a large number of the organic farmers in our case studies were farming successfully."[2] But most conventional farmers were suspicious of organic farmers and unfamiliar with their approaches. The USDA could perform a real service by funding research to fill this information gap.

We were proposing the testing of whole new ways to grow crops and raise livestock, new "systems" that simply didn't need a lot of purchased inputs—chemicals, fertilizers, fuel—to be productive.

Despite the ongoing crisis in agriculture that begged for new approaches, the LISA program was not an easy sell. Since World War II, financially successful (and unsuccessful) farmers had done the exact opposite of what I and others were now proposing. Farmers had ridden the tide of cheap energy prices and had poured on the fertilizer and pesticides. Because low-input agriculture bucked the status quo so completely, there were many misconceptions about it: It was a return to prewar agriculture, to backbreaking labor and drudgery, with low yields and low productivity.

Supporters of LISA were proposing no such thing; that was why accurate information and research were so urgently needed, and why I was willing to make the long trek from Poteau to Washington, DC. This trip to the Capitol was to be my first lesson in how government works. I was also to learn how much I and other proponents of alternative agriculture, were up against.

Testimony

The scene in front of Whitten's committee was like something most of us have seen on television: Representatives were seated at a semicircular table at the front of the room above the witnesses and audience. I seem to remember a room with a lot of heavy, dark wood. I wore a suit that day, though, as it turned out, wearing overalls might have made more of an impression. I sat at a long table before the committee and spoke into a microphone. I was scheduled to speak for about ten minutes, and as I did I was shocked by what I perceived as the indifference of Whitten and other committee members. I told them that the clock was ticking; the inescapable petroleum shortages of the future could cripple conventional farming systems, and we needed to look into alternatives without delay. Although some of the congressmen asked me questions in what I judged to be an attempt to draw the attention of their fellow committee members, in general I felt like an eager teacher in front of a room full of bored teenagers.

I left the room discouraged. I had testified in favor of a program that I believed was urgently needed to address one of the most important issues of our time: How we are to raise enough food to feed a growing population and yet preserve natural resources, environmental quality, and quality of life on the farm, all at the same time. Yet no one seemed interested. I felt cynical about the process; I speculated that, while I spoke, those congressmen on the committee had been contemplating how much agribusiness had contributed to their reelection campaigns. I already knew that the USDA had resisted setting up the program I was advocating. While the Senate had instructed them to pursue alternatives in agriculture a full three years previously, the USDA had not requested funding for such programs. Those mile-long rows of cotton I had hoed as a kid looked easy next to this hard row of bureaucracy I had decided to tackle.

Though at times I had felt alone in that committee room, I really wasn't. Others of like mind had testified with me that day—university researchers and ordinary farmers who were in favor of change. One farmer, Jess C. Andrew III of West Point, Indiana, testified that he had attempted for eight years to obtain "up-to-date, reliable information" on low-input farming systems from his land-grant university. What he had gotten instead, he felt, was completely out of sync with current agricultural realities. "Our farm, given the current state of the farm economy, needs information targeted specifically to the reduction of production costs rather than information oriented toward higher yields," he stated.[3]

Brian Chabot, associate director of research at Cornell University Agriculture College, acknowledged the importance of USDA programs that fund basic research. But he added, "We need programs that deal with real problems at the farm level. . . ."[4] As for LISA's "lack of high-tech glamour," Chabot said, "You will not be seeing fancy laboratories with expensive equipment and people in white lab coats. But this does not diminish the importance of what needs to be accomplished. There should be no less glory for those who must get their hands dirty and their boots muddy in order to help the farmers of this country."[5]

Others joined in the chorus for LISA. Michael Duffy, an extension economist at Iowa State University, testified that low-input agriculture research is vital. Donn and Susan Klor, Illinois farmers, agreed, and sketched out the changes they had made on their own 650-acre farm. Although they drastically cut herbicide applications on their corn (and thus costs), the Klors had maintained good production totals.

Charles A. Francis, extension crops specialist and professor at the University of Nebraska, reinforced the positive results of the Klors, stating, "One, it is possible to cut production costs in a number of ways, including reduced fertilizer and chemical pesticide applications; two, this will not necessarily reduce yields per acre if the right changes are made; and three, there will be a corresponding increase in net income if the right management combination is used for each crop and system."[6]

Those testifying were representative of those around the country convinced that the federal government should be focusing on what would eventually come to be known simply as sustainable agriculture. What is sustainable agriculture? It is, in large part, a wedding of ecology and agriculture. It proposes that a farm must be not only economically but also ecologically healthy if it is to be viable over the long term. It also proposes that farmers and farm workers should be afforded a good quality of life and be treated fairly. A sustainable agriculture will yield healthy rural communities and towns, which are key to the overall health of our nation.

Although I saw sustainable agriculture as a commonsense response to the problems of industrial agriculture, I knew it seemed revolutionary when measured against the agricultural paradigms of the previous forty years.

This perception hadn't scared everybody off. Those interested in this new kind of agriculture ran the gamut—farmers and ranchers of all types, environmentalists, sympathetic USDA employees, conservationists, academics, and people like me, from private nonprofit organizations. We were diverse in race and gender. I had been impressed by the dedication, tenacity, and intelligence of those in the sustainable agriculture "movement." Not surprisingly, they had much in common with environmentalists. Both tended to take the long view, believe in the value of nature, appreciate clean water and air, and have a desire to leave a better, healthier world for their children. But those interested in sustainable agriculture had an extra dimension. They understood the serious financial problems of farmers and the kinds of changes and compromises it would take to move large numbers of farmers towards sustainability. They understood that creating a sustainable agriculture is really much harder than preserving, say, a tract of old-growth forest.

What is sustainable agriculture? It is, in large part, a wedding of ecology and agriculture.

Besides feeling supported by such folks, I also took solace in the knowledge that we were just the latest in a line of agriculturists who had taken the road less traveled and who had steadfastly refused to be quiet, even when considered crackpots by the mainstream. These fore-farmers knew, as we did, the importance of agriculture—how civilizations from the Sumerians to the Mayans have risen upon its bounty and fallen upon its corruption. And how we, seemingly in the midst of plenty, were in danger of falling ourselves.

NATURE VERSUS INDUSTRY

Too often the history of agriculture is viewed from an urban perspective—emphasizing how advances in agriculture have allowed cities to grow and have "freed" farmers to move to the cities for urban work. But viewed from a rural perspective, the whole history of agriculture might best be considered as one long quest, with numerous successes and failures, to farm better.

Humans likely started farming accidentally about 9,000 to 10,000 years ago, after they had observed the sprouting of wild seeds in their garbage pits. Soon they began planting favorite wild foods near their villages. At around the same time they began to tame certain animals and raise them in captivity. Agriculture arose independently in different areas of the world. Some scholars have identified nine primeval farming areas, including one in North America that included parts of Arkansas, Kentucky, and Missouri.[7]

With the rise of farming came farming tools. Wood and stone tools such as digging sticks and sickles wielded by early farmers eventually were supplanted in some areas by wooden plows, first pulled by humans and, by 3,000 B.C. in Egypt and Mesopotamia, by oxen. (The much faster horse supplanted the ox as the the premier draft animal when new harnesses were introduced into Europe during the Middle Ages.) The plow was so important that some cultures saw the Big Dipper, one of the biggest and brightest constellations in the northern sky, as a plow.

In their desire to farm better, early farmers experimented with their crops, selecting for plants with valuable characteristics—better yield, bigger seeds, adaptation to climate. Some crops began to be utterly dependent on the farmer. Corn, as it evolved under the Native American's guidance, became a plant that cannot survive in the wild and must be planted by human hands. The seeds of valuable plants were traded and soon these crops spread over large areas. Early experiments with breeding livestock resulted in the Holstein, the first dairy cow, in 100 B.C.

With each innovation, agriculture grew more productive. In order to feed developing cities, farming became more intensive in order to yield a surplus. The results of this change were mixed, and the consequences are instructive. Around the Mediterranean Sea region, overgrazing by large flocks of sheep and goats led to erosion of the shallow soils; the grasslands turned to wasteland and remain so today. Soil erosion was a problem for many ancient civilizations, including Persia, Greece, and Rome. Even the Chinese, notable for the longevity of their agriculture, have had severe erosion problems in some areas. Mesopotamian fields were ruined by a buildup of salt in the soil from irrigation, a problem in irrigated areas yet today.

Attempting to solve these problems, past civilizations developed the first conservation practices. Phoenicians developed terracing to prevent soil erosion about 1000 B.C. The Greeks and then the Romans developed crop rotations—where plants such as beans and peas (legumes), which add nitrogen to the soil, are planted in rotation with crops that can use this nitrogen. The Romans also developed the practice of leaving half of every field fallow (unplanted) each year to store nutrients and moisture for the following year's crops, which is something my father did on our family's farm.

The Romans introduced these practices to Europe, where they were widely used and expanded. Meanwhile, in the New World, Americans had developed their own agricultural systems. In North America, particularly, where population density was rather low, Indian agriculture seems to have caused little degradation of resources. With the arrival of European explorers in the Americas, New World crops, such as potatoes, cocoa beans, peanuts, squash, sweet potatoes, tobacco, tomatoes, and corn, spread quickly around the world. European settlers learned from Native Americans how to grow these unfamiliar crops.

Tobacco became an important crop for settlers in what would become the United States, especially in the South. However, the way the Southern planters approached the raising of tobacco had little in common with the Indian approach. In his book about the Lewis and Clark expedition, *Undaunted Courage,* the historian Stephen Ambrose has characterized tobacco culture as "an all-out assault on the environment for the sake of a crop that did no good and much harm to people's health as well as to the land. . . ." He says gentlemen planters such as Meriwether Lewis "made no use of animal manure, and practiced only the most rudimentary crop rotation."[8] When the land was exhausted, which it was quickly, the planters moved to virgin land (which they were always in need of) and had their slaves begin the same process.

In contrast, German immigrants in the Shenandoah Valley kept small-land holdings, manured their fields, and rotated their crops. No slaves or indentured servants—"men with little interest in the precious undertaking of making a family farm,"[9] as Ambrose puts it—were responsible for the stewardship of the farm. These were family farms, worked by the family and consequently cared for by the family.

Lewis was ruining his land at the same time as what is now known as the Agricultural Revolution was spreading from Great Britain to the rest of Europe and North America. The revolution was multifaceted. First, a new crop rotation system made land much more productive. It was based on knowledge of which crops added or took which nutrients from the soil. Advances also occurred in livestock breeding. And, perhaps most significant, inventors developed myriad new farm machinery: the first seed drill, the first cotton gin, the first reaper, and, last but not least, the steel plow, invented by John Deere, an Illinois blacksmith, in 1837. This last enabled the eventual plowing of the grassy, rich prairie soils of the American Midwest.

In 1862, Congress passed two bills that greatly affected the future of agriculture in this country. One was the Homestead Act, which granted 160 acres of land to anyone who lived on it for five years and developed it for farming. After the Civil War, this act helped to populate the Great Plains, turning it into a major wheat-producing area. Ironically, this plowing later led to the dust bowl and mass depopulation of the area. Congress also passed the Morrill Act in 1862, which led to the founding of the first land-grant colleges, which stressed agriculture and the mechanical arts. (In 1890, land-grant colleges specifically for African Americans were established.)

In the early decades of the twentieth century, agriculture began to become modernized. Gasoline- and then diesel-powered tractors began slowly to replace the horse and mule. Some call this the beginning of the Mechanical Era.

It was also the time, says Richard Harwood in *Sustainable Agricultural Systems,* when farmers established organizations to "develop and share technological knowledge among themselves." If it was the beginning of the information age in agriculture, it was also the beginning of a split in agriculture, a split still evident that spring day in Washington when I testified before that House committee. Some farmers' groups looked to industry for information and a model. Others looked to nature.[10]

The Industrial Model

The farmers who adopted the industrial model created industrial agriculture, which is the agriculture we have today. In the latter half of the twentieth century, "better farming" the industrial way has meant increasing production and increasing productivity. Low-cost energy converted into cheap fuel for bigger tractors, and myriad fertilizers and pesticides made possible massive increases in production: Farmers in 1930 fed 9.8 persons in the United States and abroad; today, as noted earlier, 129. Figures on commercial fertilizer use help tell the tale: from about four million tons used annually in the first decade of the century to over twenty million tons used annually in the last decade of the century.

These spectacular increases in production were also made possible through specialization (planting just one or two crops), using ever bigger and more powerful machinery, and planting higher-yielding hybrid crops often bred to respond to the use of chemical fertilizer. (Use of these hybrids created the green revolution in the third world.) The industrial ideals of specialization, efficiency, and uniformity of product were applied to agriculture. Average acreages got bigger, loans got bigger, all the numbers got bigger, it seemed, but the number of farmers—from 38 percent of the labor force in 1900 to around 2.5 percent by century's end. Farming as a way of life was not possible anymore for many who wanted it; it required too much money and too much land.

The land-grant colleges encouraged the use of industrial farming practices. Not only did these schools educate most agriculture students, they each also had an agricultural experiment station, laboratories, and experimental farms. The cooperative extension system, a partnership between the federal, state, and county governments, distributed information from the land-grant universities to farmers. County agricultural agents "demonstrated" new practices for farmers to follow.

As Harwood points out: "By the late 1950s, the evolution and spread of industrial technologies had increased exponentially. In the developed nations, the industrial model was widespread."[11]

Some farmers' groups looked to industry for information and a model. Others looked to nature.

The Nature Model

Although industrial agriculture promised its practitioners and society a prosperous future, others were not convinced it would do that. Biological agriculture is a good name for the various "schools" of farmers who looked to nature rather than industry as a model of how a farm should work. Their ideas developed parallel to those of industrial farmers, and offered an alternative model of how best to farm.[12]

Many of the ideas explored by the biological farmers are not new. As far back as Lincoln's presidency, green manures (a crop planted to turn under and enrich the soil) were a hot topic in agriculture. Leading thinkers of the day were interested in soil health—Charles Darwin wrote about earthworms and their effects on soil. Rotating crops, adding animal and green manures, and proper tillage were of primary importance.

In the 1920s, Rudolf Steiner, an Austrian philosopher, gave a series of lectures on what he called biodynamic farming. Many biodynamic tenets stood in stark contrast to what would become conventional: planting a diversity of crops rather than one or two, recycling waste, avoiding chemicals, and decentralizing production and distribution. Although Steiner also espoused more esoteric beliefs, such as the effect of cosmic forces on plant growth, his other tenets were shared by the humus and then the organic schools of agriculture.

In contrast to the conventional approach, these schools of agriculture focused on the health of the soil and emphasized farming techniques that improved soil structure and fertility. A leading advocate in the humus school was Sir Albert Howard, a British agriculturist working in India. Howard said: "The maintenance of the fertility of the soil is the first condition of any permanent system of agriculture. In the ordinary processes of crop production fertility is steadily lost: its continuous restoration by means of manuring and soil management is therefore imperative."[13]

Some advocates of the biological approach were land-grant scientists, while others were farmers, or independent thinkers such as J. I. Rodale, the founder of Rodale Press, publisher of the magazines *Organic Gardening* and *Prevention,* who wrote books on organic farming in the 1940s and 1950s. Influenced by Howard, Rodale and others eschewed chemical fertilizer, believing it to have detrimental effects on the soil. Even the plow was reassessed: Edward H. Faulkner argued in *Plowman's Folly* that the moldboard plow was detrimental to soil health because it left the plowed soil vulnerable to erosion.[14]

These farmers did not rule out every aspect of conventional farming—they too adopted new crop varieties, farm machinery, and soil nutrient testing.[15]

These variously named yet similar approaches to farming were in the minority. By the time I went to college in 1965, as Harwood points out, industrial agriculture seemed so successful that alternatives were not seriously debated. The schools of alternative agriculture

> were practically nonexistent and certainly in disrepute . . . gone were the traditions of humus farming, of mechanical weed control, and of the need for large portions of our population to be involved

> in agriculture. Farming was now a business, to be run as efficiently as any other industrial enterprise.[16]

I attended Oklahoma State University, formerly Oklahoma A and M, Oklahoma's land-grant college. Looking back, I believe I got a good, if narrow, education there. The professors cared about their students. Courses were intense and relevant, addressing many of the production problems I had seen on my father's farm. I acquired a belief in the importance of scientific research to agriculture, and the importance of extension work—of passing along the results of that research to farmers in the field not only in Oklahoma but around the world. We believed these were sunny days for agriculture. American farmers were showing the rest of the world how to farm the industrial way. We had entered the Chemical Era, and we were happy about it.

Not surprisingly, given the apparent success of the accepted approaches, some ideas were not discussed in the classroom. In college, I heard about organic farming, but it wasn't part of the curriculum. Environmental ideas were just as rare and perhaps even more scorned. Rachel Carson's book *Silent Spring,* warning of the dangers of pesticides such as DDT, appeared in 1962. I hadn't read it, but in class one day I asked a professor I respected, an entomologist, what he thought of Carson's ideas. He replied that she was a lunatic. That ended the discussion. Such thinking was common in agricultural circles, though Carson turned out to be right. The public became ever more concerned about the negative effects of pesticides on wildlife, and the presence of pesticide residues in food, while industrial agriculture resisted change.

The organization I now head, the Kerr Center for Sustainable Agriculture, had its roots in industrial agriculture. Its predecessor was the agricultural division of the nonprofit Kerr Foundation, founded in 1965. Named for the late Oklahoma senator Robert S. Kerr, a cattle rancher and ardent conservationist, established on his land, and funded by his estate, the organization stressed conventional agricultural solutions along with conservation practices advocated by the U.S. Department of Agriculture's Soil Conservation Service.

But even given this belief in conservation, at the "ag" division certain ideas were taboo. I occasionally bought a copy of *Organic Farming and Gardening* at a newsstand. I thought it was interesting, but would never have subscribed to it, or admitted to my colleagues that I read it. It was the kind of magazine one kept hidden—like *Playboy.*

The Rodales we dismissed as health fanatics, distrusted in part, I think, because they were from Pennsylvania, "Back East." They weren't "true farmers," we thought.

Thinking back on it, we were arrogant. We were dazzled by our own success. We didn't consider that when the industrial agriculture of the past fifty years was placed in the 10,000-year history of agriculture, it was a new, rather experimental development. Of course, we merely reflected the tenor of the times, when most people believed that there would be a chemical fix for everything: bad soil—put more fertilizer down; insects—spray insecticide and often.

Rethinking Industrial Agriculture

My first inklings that the industrial agriculture system was not all it was cracked up to be came gradually, as I worked with small farmers and realized how difficult it was for them to stay in business. I had grown up happy on a small farm, and I believed it was a worthwhile way of life—a way of life that should at least be an option for people in rural Oklahoma.

We didn't consider that when the industrial agriculture of the past fifty years was placed in the 10,000-year history of agriculture, it was a new, rather experimental development.

But small farms were disappearing, unable to compete. One by one, the families in my native community south of Cold Springs left because larger farms outbid them on the rented land they used. Such situations led me to reason that small farms should not try to compete with big spreads, but find a profitable niche of their own. What I discovered, however, was that farmers operating small farms, who needed the most help, were often the last ones to get it from extension agents and government programs. Information was geared to those who wanted to grow what everyone else grew—wheat, corn, and soybeans—on large acreages rather than alternative crops on small ones.

As for loans: it was hard to get credit for small farm enterprises from either the Farmers Home Administration (FmHA) or commercial banks. Although it was true that one could always borrow money for tried and true enterprises such as cattle, credit was difficult to get for something new, such as truck farming. Such enterprises were out of the lender's comfort zone. As one catfish farmer complained to me when he couldn't get a loan: "They couldn't count my collateral." Unfortunately for him, his "livestock" did not have four legs and say moo.

The ag division was located in a county of mostly small farms and ranches, and we began to try and fill some information gaps. From 1975 to 1979 we hosted a number of small farm conferences, trying to create an awareness of problems. We worked to get these forgotten farmers information on alternative enterprises, and any government programs that could help them. I worked with Frank Evans, a top FmHA official in Oklahoma, in an effort to educate lenders and, much to my satisfaction, they responded. It was my first taste of how the USDA works, and I was gratified to find that change can occur when individuals take it upon themselves to make it happen.

Lenders began to refer to me those wanting to get loans for alternative enterprises, so I could "shepherd" them through the process. The FmHA required a "whole farm plan" before lending to unconventional enterprises, and I helped farmers prepare one. My goal was to get these farmers a fair hearing.

At around the same time I became an advocate for small farmers, others around the country were becoming advocates for a cleaner environment and turned their attention to the environmental problems being caused by industrial agriculture. Because some of these advocates were not farmers or part of the agricultural establishment, their interest was, to put it mildly, not welcomed. And because sometimes these advocates were rather romantic and not well versed in the economics of conventional agriculture, they were easy targets for those who didn't want to address what turned out to be legitimate concerns.

The silliest kind of rhetoric was hauled out to attack them—for instance, organic farmers were accused of wanting to take us back to "a Tarzan life among the apes." This was asserted in the pamphlet, *Facts from Our Environment,* published by the Phosphate Institute (phosphate is a key ingredient in chemical fertilizer). About 150,000 copies were printed and distributed through institutions such as the Kerr Foundation.[17]

The response of some in industrial agriculture to organic farmers brings to mind the line, paraphrased, from Shakespeare's *Hamlet:* "Methinks they doth protest too much." False dichotomies were set up—in Nebraska "an experiment" was conducted comparing corn grown with conventional inputs and corn grown without. The corn grown with the inputs, dubbed "Today's Acre," vastly outproduced the other, called "Nature's Acre," and this was served up as proof that if inputs were cut, the world would starve.

Of course, the corn grown without inputs was grown in soil that had been depleted by years of farming with chemicals, and had no fertility left: It was bound to fail. The field in no way resembled the fields of organic farmers who painstakingly build up their soil with organic matter and organic fertilizers so that their land is fertile. Ironically, while the "experimenters" didn't plan it that way, the experiment more accurately reflected just how worn out and dependent on chemicals the conventional agricultural fields were, buttressing the argument that something indeed needed to change.

Public concern about agriculture in the long run could not be ridiculed out of existence. In the 1980 *Report and Recommendations on Organic Farming,* the USDA noted the concerns of the general public, farmers, and environmental groups about the country's system of food production. The study noted found that people were concerned, as I was, over the demise of the family farm and localized marketing systems. Other concerns included the declining quality of the nation's soils, farmer dependence on energy and agrichemicals and their rising costs, pollution of surface water and groundwater by common agrichemicals, and pesticide residues on food and their possible effects. Consequently, the study stated, "many feel that a shift to some degree from conventional (that is, chemical-intensive) toward organic farming would alleviate some of these adverse effects, and in the long term would ensure a more stable, sustainable, and profitable agricultural system."[18]

It seems that the public had noticed the excesses of industrial agriculture as outlined in my indictments. And they had also grasped a possible solution. The study found that a small group of farmers who had switched to a low-purchased input approach were successful—environmentally sound, energy conserving, yet stable, productive, and profitable—and, over the long term, sustainable. Although noting some problems, the report concluded that much could be learned by further investigating organic farming methods. The biological farmers had at last been given some respect.

The study was a watershed, and indeed some consider 1980 as the end of the Chemical Era in agriculture and the beginning of the Sustainable Era.

THE SUSTAINABLE ERA

As former Environmental Protection Agency director William Ruckelshaus has observed, "sustainability was the original economy of our species."[19] Preindustrial societies had to sustain the ecosystems upon which they depended for food, shelter, and warmth, for their only alternatives were to move or to perish.[20] With technology, our alternatives have expanded, obscuring the fact that the bottom line is still the same: sustain the ecosystem or pay the consequences sooner or later.

It took me awhile to really believe this, to get this on a gut level. I originally came to sustainable agriculture not through concern about the environment, but through concern about the small farmer. In the early 1980s, after I became director of the ag division, we worked on lowering input costs on our 4,000-acre ranch. We planted clovers and other legumes in the pasture sod to reduce nitrogen fertilizer costs, and, by spraying weeds earlier, used less herbicides. Information about these changes we passed on to area farmers who needed information about cutting costs.

As far as the environment was concerned, I believed agriculture would have to address groundwater pollution and soil erosion more completely someday. But before I began learning about sustainable agriculture, I didn't have the faintest idea about how to do it. Revelations about the health dangers of Agent Orange gave me pause; two of its ingredients, the herbicides 2,4,5-T and 2,4-D, had been used by farmers. As far as the environmental movement was concerned, I really thought it would never amount to a hill of beans. I hadn't yet grasped that environmental risks as well as financial risks were a threat to agricultural sustainability.

Clearly, I've come a long way. My thinking about sustainable agriculture changed gradually. I did not have a blinding insight one day that changed my mind forever; instead, I had many changes of heart.

My education in sustainable agriculture began in a series of conversations I had with a member of the Kerr Foundation's board of trustees, Kay Kerr Adair, the daughter of the late senator. She came from outside of agriculture, yet had a keen interest in it. She had no agenda, and she was not a captive of a vested interest; she had simply read some books that caused her to think that something was wrong in conventional agriculture. Whenever we met, she would ask me questions: What do you think is happening on the land? Are we really helping people?

I did not have a blinding insight one day that changed my mind forever; instead, I had many changes of heart.

Although these were not easy questions to answer, I found her honest questioning to be refreshing. I also found refreshing her abiding spiritual belief that we need to care for the earth. She thought that, on a farm practicing sustainable agriculture, it should be possible for the farm family to stay in touch with nature, an essential for spiritual health.

On the other hand, I was concerned that she didn't have any farming experience. I countered her with the standard argument: agriculture can't change, because the world will starve if it does. She didn't demand that I change my ideas, but our first brief conversations caused me to at least start thinking about the basic premises of conventional agriculture.

At about this time, the agriculture division's parent organization, the Kerr Foundation, underwent a reorganization. In order for its charitable efforts to be more focused, the foundation was divided into four separate entities, with each new foundation to be guided by one of the four Kerr children. Kay Kerr Adair was to head the board of trustees of a new agriculture foundation that would take the place of the agricultural division. This new foundation, although retaining the educational focus of the ag division, would be oriented toward alternative agriculture.

In May of 1985, a new board of trustees formed to guide the new, still nonprofit foundation. A new name had to be decided upon: Would it be the Kerr Center for Alternative Agriculture? Regenerative Agriculture? Organic Agriculture? Not only was the name up in the air, but, so it turned out, were the feelings of my staff—half of them quit right after the new organization was named the Kerr Center for Sustainable Agriculture.

I understood their fears. I wasn't sure myself if I could make the transition. Some new board members made me very nervous because I considered them radical environmentalists. I found myself meeting often with Kay and her husband Robert C. Adair Sr. to exchange ideas. The discussions about what to do at the new Kerr Center were intense. A lot was happening in sustainable agriculture and I was exposed to many new ideas.

Often sustainable agriculture offered quite a different approach to solving farm problems than I was used to. On the long drives home after visiting the Adairs, I thought about how to explain these ideas to my staff and, more to the point, I tried to make sense of them myself.

The transition was complicated when I was almost immediately faced with a big decision. The governor of Oklahoma, Henry Bellmon, offered me the job of secretary of agriculture. I had a choice: climb the final rung to the top of ag ladder in Oklahoma or stay in a difficult job I was not sure I could do.

I picked the difficult job. Why? It was partly because I didn't want to leave my farm and my community; partly because I worried about the political nature of the secretary's job as well as its brevity. But, most important, I stayed because I wanted the challenge—I was beginning to feel that if I could find a way to answer my own questions about sustainable agriculture, I could make an important contribution to Oklahoma's agriculture. I was no radical, but I had farm experience, and I hoped that I could help farmers make what I believed would be a long slow change to a sustainable agriculture.

The transition time at the Kerr Center was also complicated by the reaction of my peers in agriculture in Oklahoma. I served on a lot of advisory boards and committees, and though no one actually said anything negative to me, no one asked me any questions about what we were doing, either. I think they were embarrassed for me—as if I had been arrested for drunk driving or some other shameful kind of incident, ruining my brilliant future. To say the least, sustainable agriculture was not a popular concept in Oklahoma—not surprising, given the extreme reaction of many in conventional agriculture; people did not have an accurate idea of what sustainable agriculture was all about. A lot of antienvironmentalist feelings existed in the state, too, and I felt like people put me in that category. No longer did I have the comfortable role of trusted adviser, helping the farmer or rancher calibrate a sprayer or get a loan. So it took a long while for me to get comfortable—to figure out for myself what sustainable agriculture meant—and to be able to acknowledge that, yes, sustainable agriculture is about protecting the environment. It's also about helping the farmer make it financially, and about preserving natural resources and quality of life in rural communities, and about everything that agriculture should be about.

I was no radical, but I had farm experience, and I hoped that I could help farmers make what I believed would be a long slow change to a sustainable agriculture.

David and Goliath

While I was going through my personal struggles, and the Kerr Center was trying to find its mission, the United States government first formally recognized sustainable agriculture in the 1985 Farm Bill. The Agricultural Productivity Act authorized the USDA to look into research and education in low-input sustainable agriculture.

But it hadn't happened, which brings us back to April of 1987 when I went to Washington. The day after I testified to the House of Representatives in support of LISA, I gave the same testimony to the Senate Appropriations Subcommittee on Agriculture, Rural Development, and Related Agencies chaired by Quentin N. Burdick.

I realized just how much my life had changed when I was joined in my testimony by *Organic Gardening* magazine publisher Robert Rodale, son of organic farming pioneer J. I. Rodale. He spoke of the on-farm research and demonstration network the Rodale Institute had recently established, and the conference on low-input farming they had cosponsored earlier that year. He expressed support for what he called "regenerative" farmers—those who improve natural resources, such as soil, while using them to grow crops.

I shared Robert Rodale's belief that farmers can be builders, not just takers. So, I guess, did some committee members, because our testimony had an impact. Despite my pessimism—and over the objections of the Reagan administration—later that year Congress appropriated 3.9 million dollars to establish LISA, finally funding the research and education called for in the 1985 Farm Bill. It was a far cry from the thirty million dollars we had hoped for, but it was a start—seed money. I don't think I am far off in saying the action marked a small triumph of representative democracy. If 1980 marked the beginning of the Sustainable Era, this was its first milestone.

On his last day in office, then-Secretary of Agriculture Edward Madigan issued a memorandum establishing departmental policy regarding alternative (now termed low-input/sustainable) farming systems. The Department of Agriculture officially encouraged research and education that would provide farmers with information about farming systems which minimized the use of costly and environmentally hazardous inputs.

This statement was a victory for sustainable agriculture. But, as it turned out, the war was far from over. There were to be more skirmishes ahead. The first of these took place just a month later when, in a press release announcing the LISA program, the USDA's Assistant Secretary of Science and Education Orville Bentley praised low-input farming with its more prudent use of pesticides and commercial fertilizer as an idea whose time had come. That got the attention of the Fertilizer Institute, a trade organization. The group objected to the statements and, in a letter to the USDA, accused the agency of losing touch with the American farmer by promoting "cockamamie" low-input ideas. Indeed, the very name LISA—low-input sustainable agriculture—seemed to act as a call to arms for such groups.

Their distress was not surprising. In 1985, the fertilizer industry had sold millions of tons of fertilizer to American farmers. A program that looked at ways to decrease fertilizer use would naturally be seen as a threat. What happened next, though, was a bit surprising. The Deputy Secretary of Agriculture, Peter Myers, reportedly apologized to the head of the Fertilizer Institute for the announcement's "hard-hitting" language. And what's more, the USDA let the Fertilizer Institute review a draft of Bentley's response to its protest, which was then reportedly discarded and a milder version substituted.[21]

These exchanges were revealed during another Congressional hearing the next year. Chaired by Oklahoma Congressman Mike Synar of the Government Operations Subcommittee on Environment, Energy, and Natural Resources, the hearing looked into how the USDA and the EPA were dealing with agrichemical pollution in groundwater. "There are clearly some mixed signals," Synar told Myers. "We're [the USDA] talking a good story here, but attitude and commitment are not backing it up."[22]

Despite its run-in with the Fertilizer Institute, the USDA did implement the LISA program. A diverse group of highly motivated agriculturists both within the USDA and from nonprofit organizations designed LISA to be science-based, grass roots, and oriented toward problem solving. It reached beyond the USDA to involve farmers, ranchers, and nonprofit organizations in policy development, management, and oversight, and in the technical review of proposed projects. This democratic, inclusive emphasis was unique. Extension programs would be used to communicate research findings.[23]

LISA was followed by a more comprehensive and targeted set of policies and program goals found in the Food, Agriculture, Conservation, and Trade Act of 1990 (FACTA). In this bill, the Sustainable Agriculture Research and Education (SARE) program was established to replace the former LISA program. The programs were basically the same, but the name SARE (with the term *low-input* removed) represented less of a lightning rod to opposition groups.

Defining Sustainable Agriculture

Sustainable agriculture was defined in the Food, Agriculture, Conservation, and Trade Act (FACTA) of 1990 as:

an integrated system of plant and animal production practices having a site-specific application that will, over the long term:

(a) satisfy human food and fiber needs;
(b) enhance environmental quality and the natural resource base upon which the agriculture economy depends;
(c) make the most efficient use of nonrenewable resources and on-farm resources and integrate, where appropriate, natural biological cycles and controls;
(d) sustain the economic vitality of farm operations; and
(e) enhance the quality of life for farmers and society as a whole.

Source: U.S. Congress, 1990. Title XVI, Research, Subtitle A. Section 1602.

SARE is a scrappy little program without a lot of money but with a "noble purpose."[24] There are four regional offices associated with a host university. Each year the four regions (Northeast, North Central, Southern, and Western) fund a gamut of research projects that will help make agriculture sustainable. Since 1988, the program has funded close to 1,200 projects in three categories: research and education grants typically given to universities or nonprofit organizations often in cooperation with farmers; producer grants which go to farmers and ranchers for on-farm research; and professional development grants which fund training in sustainable agriculture practices and concepts for agriculture professionals such as extension agents and specialists with the Natural Resources Conservation Service.

There are many good things to say about SARE as a program. Its overhead is low: 94 percent of its funding goes to research and education. It's not a Washington bureaucracy; decisions on how money is spent are made on the regional level. Members of technical committees and administrative councils who make such decisions are local and diverse. Indeed, local supporters have matched every federal dollar spent with eighty cents. The projects funded are relevant to life as we know it: farmers and ranchers participate at all levels, ensuring the utility and practicality of projects. In fact, the program has been praised by the General Accounting Office for having ". . . successfully involved often opposing entities, including farmers, nonprofit organizations, agribusiness, and public and private research and extension institutions."[25]

Like the trend towards giving some federal power back to states, sustainable agriculture chips away at the monolith of the federal agriculture establishment and transfers power and money and expertise back to local areas. For several years I have been involved with the Southern Region of SARE. I have often sat in on meetings where a committee debates the merits of proposed research projects. Each committee is composed of a mix of university researchers, farmers, and representatives of nonprofit organizations similar to the Kerr Center. These are folks who, if it were not for SARE, would probably never have met. The exchange that goes on in such meetings and in SARE projects helps keep the program well rooted. Watching as a committee eventually comes to a consensus, I can't help but think that if this process could occur more often, then American agriculture might really become the greatest system in the world.

The projects funded are relevant to life as we know it: farmers and ranchers participate at all levels, ensuring the utility and practicality of projects.

That said, all this praise for the structure of the program begs the question: Does the program deliver? Do SARE projects demonstrate that alternative approaches to raising and marketing crops and animals work?

The evidence that they do are the regional project updates published yearly. A few highlights: Cranberry farmers in the Northeast have cut herbicide, insecticide, and fertilizer use by about half. Researchers in Florida and Texas found organic pest-control strategies could save growers about $400 per acre and reduce workers' exposure to pesticides. Farmers in Iowa have found that a new way to till the soil, called ridge tillage, controlled erosion and reduced inputs. Fruit and vegetable producers in Connecticut are supplying Hartford schools with fresh produce; farmers gained a new market, and inner-city students not only got high quality fresh food, but learned about agriculture, nutrition and the environment at farm field days and chef demonstrations.

I think it is not too rash to say that, with the establishment of LISA/SARE, the Goliath of industrial agriculture has finally met its David.

And as for alternative agriculture being "a Tarzan life among the apes," I beg to differ. These projects, rather, represent real progress. There are hundreds of such innovations going on around the United States and thousands more around the world, done by myriad organizations and individuals. The search for a healthy, enduring agriculture—a sustainable agriculture—is on.

WHAT IS SUSTAINABLE?

"Sustainable" is usually defined as "enduring." We are really just beginning to identify what might make a sustainable agriculture. Presently, what is thought of as sustainable agriculture encompasses a variety of philosophies and farming techniques—in general, these are low chemical, resource and energy conserving, and resource efficient.[26]

Sustainable farms are more self-sufficient than conventional ones, relying less on banks and government subsidies. This independence is often overlooked, but is important in this era of dwindling federal support for agriculture. According to the National Research Council in a 1989 report, "Farmers who adopt alternative farming systems often have productive and profitable operations, even though these farms usually function with relatively little help from commodity income and price support programs or Extension."[27]

Agroecologists attempt to learn from natural ecosystems, as ecologists do, and apply what they learn to agriculture.

The philosophical basis for sustainable agriculture, as I noted earlier, is with the biological agriculturists of the earlier twentieth century. More recently, inspiration has come from the field of agroecology. Agroecologists attempt to learn from natural ecosystems, as ecologists do, and apply what they learn to agriculture. Agroecologists view the farm as a particular kind of ecosystem, an agroecosystem. They view the farm "system" as a whole, with the farmer and farm family as one part of the whole. All the parts of a farm are interrelated. The key is to create as many natural balances in the system as possible. To be sustainable, say agroecologists, a farm should be modeled as much as possible after a healthy natural ecosystem, such as a forest or prairie.

What does it mean to model a farm after a forest? Like a natural ecosystem, a sustainable agroecosystem is powered by the sun through photosynthesis; it generates much or all of its own fertility and pest resistance through complementary interaction among plants, animals, and soil organisms; it contains a wide variety of species of plants and animals adapted to local conditions of climate and soil type.

In contrast, conventional agroecosystems depend heavily on inputs from the outside to be maintained—thus fertility is supplied by purchased chemical fertilizer instead of by animal manure from animals on the farm or by legumes grown on the farm; pest control is accomplished with purchased pesticides instead of with beneficial insects which live on the farm.

A natural ecosystem sustains itself without human interference. This may not be possible in an agroecosystem because so much is removed from it each year in the form of harvested crops or milk or beef. But the goal is to move it toward self-maintenance.

This "systems approach" is key and can be broadened. The system can be expanded beyond the farm to include the local ecosystem of forests, rivers, or other natural elements, and further on to the communities affected by the farm both in the local area and globally.[28]

The concept of "sustainability" first came up during the energy crisis of the 1970s in relation to our supply of oil. It was then defined as "maintaining the present without compromising the future."[29]

In agriculture the term tackles the central challenge of the new century: making an agriculture that is economically profitable in the short term, while also working toward systems of agriculture that will be ecologically healthy in the long term. Industrial agriculture focuses almost exclusively on short-term gain without much thought about the long term. So we see dead soil, dwindling oil supplies, polluted water, and displaced farmers.

A sustainable agriculture is one that, ideally, can exist indefinitely. Who can be against this? Particularly since, as others have pointed out, "the obvious implication of an unsustainable agriculture is massive starvation and potentially the demise of the human race."[30]

A sustainable agriculture must fulfill the food needs of a growing world population. Too often, conventional agriculturists have used the population explosion as a justification for their farming methods. Again, the rhetoric has been inflammatory—as when Earl Butz, secretary of agriculture under Richard Nixon, pronounced that going back to organic agriculture would doom millions to starvation.

Butz and those who agreed with him seemed not to recognize that the world hunger problem is complex and often due to problems of waste and distribution rather than supply. One small indication of this: The USDA estimates that 27 percent of the food served in this country is wasted. At the same time, according to Tufts University, 35 million Americans are hungry or unsure about what or when their next meal might be.[31]

In the developing world, the green revolution that was supposed to solve the problem of hunger has had unintended negative effects on rural peoples. The green revolution focused on raising production to decrease poverty and hunger, and this strategy was implemented among farmers who farmed better soils, irrigated, and had substantial assets. However, this served to marginalize much of the rural population, actually decreasing their access to croplands, grazing lands, and water supplies. Worldwide, there are more than a billion farmers with very limited assets. Industrial agricultural approaches that emphasize technology are beyond the reach of these farmers.[32]

These are the very same people who most often fall victim to starvation. According to Nobel Prize-winning economist Amartya Sen, it is poverty, not absolute food shortage, that has been the primary cause of starvation in the world. Even during famine, food was often available. But those with low incomes, such as landless laborers or small farmers, had no money to buy it.[33]

Sustainable approaches can benefit these farmers, and using them does not necessarily mean a loss in yield. Farmers are proving every day that it can be done. Furthermore, restoring the health of our agroecosystems is essential if crop yields are to continue to meet food demands. Currently, yields are stabilizing or declining in most of the intensively farmed regions of the world. As agricultural economist and sustainable development expert Charles M. Benbrook has pointed out, "Achieving a higher degree of global food security will depend on reversing the decline in natural resource productivity and in enhancing *biological* productivity and resiliency of farming systems. . . . It will arise from success in restoring the *biological* integrity of soils, worldwide."[34] (emphasis mine)

. . . restoring the health of our agroecosystems is essential if crop yields are to continue to meet food demands.

Definitions

Like many complex issues in our society, a sustainable agriculture has multiple facets. Some see sustainable agriculture as a philosophy based on understanding the long-term impact of our activities on the environment and on other species.[35] This definition addresses the belief, which I share, that agriculturists must take the long view.

Another definition of sustainable agriculture I like is "an agriculture that can evolve indefinitely toward greater human utility, greater efficiency of resource use, and a balance with the environment that is favorable both to humans and to most other species."[36] This includes the idea that sustainable agriculture is not one method of agriculture, set in stone, but an agriculture that is evolving and adapting to place and circumstance.

This is an important point. Sustainable agriculture does not mandate a specific set of farming practices. There are myriad approaches to farming that may be sustainable. Because sustainable agriculture will continue to be defined farm by farm and individual by individual, it diverges sharply from industrial agriculture, which claims to be appropriate everywhere. Sustainable agriculture, on the other hand, holds that sustainable approaches will vary from site to site. For example, drought-resistant crops and water-conserving technology may help make farming sustainable in Israel and the Middle East where, in fifteen years, it is projected that there won't be enough water available for agricultural uses.[37] In eastern Oklahoma, where there is plenty of rain, sustainable agriculture may mean putting to use the mountains of local chicken litter produced by chicken farmers in a way that doesn't pollute local water.

Sometimes it is easier to understand what is sustainable by looking at what is not. On a practical level, one can identify a problem—say a turbid stream. The stream is muddy because of soil erosion from surrounding overgrazed pastures. If overgrazing pastureland causes erosion, what is the underlying cause? Is less than optimal forage being used for the class of livestock in question? Did economics force the rancher to place too many animals on a given area? Did low soil fertility contribute to a less than ideal ground cover, thus exposing the soil to the elements? All of these problems point to a system that is unsustainable.

The Leopold Center in Iowa has defined sustainable agriculture as "farming systems that are environmentally sound, profitable, productive, and maintain the social fabric of the rural community." This inclusion of the social is important; in fact, Neil Hamilton of Drake University includes in his notion of sustainable agriculture the preservation of the social values contributed by the agricultural community to U.S. society.[38]

Put simply: Sustainable agriculture includes stewardship of both natural and human resources. This includes concern over the living and working conditions of farm laborers, consumer health and safety, and the needs of rural communities.[39]

The quality of life in rural areas has been declining. A sustainable agriculture would help reverse that. This should be good news to the thousands of Americans who are moving to small towns and rural areas each year in search of a better quality of life. These refugees from urban blight certainly don't want to encounter rural blight: polluted streams, stench from industrial hog farms, and collapsed communities. Many would like to grow or buy fresh produce and contribute to a viable rural economy.

Over the years, I have drawn on all these definitions to come to my own definition. For an agriculture to be sustainable I believe it ought to be science based, farmer driven, and profitable. It should contribute to, or at least not detract from, the environmental health of the area. It must be consumer friendly, delivering safe, nutritious food. It should provide the basis for strong rural communities. The problem with industrial agriculture is that it too often emphasizes just one of these factors—profit through high production—which has left a lot out, including a lot of people.

A decade after the creation of LISA/SARE, there is still a wide divergence within the agricultural world on the best road to the future. On one end of the spectrum are the organic farmers. These are people who have worked hard to make a place for themselves in American agriculture by pioneering in the search for alternative farming methods and in the creation of new markets for food produced without chemicals. However, some of the people on this end, when it comes to defining sustainable agriculture, can be purists. Some do not see compromise, or a gradual transition from conventional methods, to be valid.

At the other end of the spectrum are those in conventional agriculture who have decided that the best defense is a good offense. They claim loudly that business as usual is, in fact, sustainable. They do not believe that erosion rates are serious. They see the trend for fewer but bigger farms as good and an economic necessity. This view is most often held by larger farmers, the individuals and agencies that helped them get bigger, and those who only look at numbers and curves. They believe in saving the world with pesticides and plastics, as one recent book proposed.[40] They don't address social issues, like the displacement of farmers from their land, and they downplay environmental problems.

Ups and Downs: The Sustainable Blues

Up: Because of continued pressure by the public and sustainable groups, the USDA's sustainable agriculture research and education program, SARE, has been funded each year.

Down: Despite the success of the program, increases in funding for SARE have been hard to get. In 2001, the SARE programs received 15 million dollars out of a total budget of over one billion dollars.

Up: In 1991 the United States Department of Agriculture and the Environmental Protection Agency cooperated in establishing a program similar to SARE. Agriculture in Concert with the Environment (ACE) focused on reducing the misuse of agricultural chemicals and animal waste, encouraging the use of biological controls and reduced-risk pesticides, and protecting ecologically sensitive areas.

Down: Since 1995, in an unfortunate budget-cutting move, the EPA reduced its contribution to the program.

Up: The USDA created a National Sustainable Agriculture Advisory Council (NSAC) in 1993. The Council was created because the General Accounting Office (GAO) had found that the USDA lacked a departmental policy to provide clear and comprehensive goals for the nine agencies involved in sustainable agriculture.[41] I was named chairman of the group, and we were to make recommendations about research and extension projects directly to the Secretary of Agriculture.

Down: The NSAC never had any real power or influence. Before we were abolished (without formal notice from above), we were underfunded (making it difficult to even get together for a meeting), our draft policy statement was ignored, and even our letterhead, which listed members of the council, was objected to!

Up: Interest in the SARE program among researchers and producers remains high.

Down: Because of a lack of funding, only a little over one-quarter of the proposals received from 1988 through 1993 could be funded. Many high-quality projects remain unexplored due to continued low budgets.

Most farmers who are trying to make their farms sustainable fall somewhere in between these two ends of the spectrum. At the Kerr Center for Sustainable Agriculture, we have tried to build a bridge between the two ends. Our approach is broad. For us, a sustainable farm is both ecologically sound and economically profitable. We are concerned about social issues as well as production issues, and we lean toward a practical philosophy with plenty of contact with and input from real farmers. We are tolerant of some chemical inputs (fertilizer, pesticides) as long as they are not used excessively. Yet, at the same time, we are always looking for cheaper, safer alternatives—ways to take advantage of nature-sponsored fertility and adapted crops. Our goal, both on our home ranch and when advising farmers, is to move a little closer each year toward being self-sufficient. We use our experiences to show farmers how to make the transition from an industrial orientation to a sustainable one.

Every organization or individual grappling with the question of what makes a sustainable agriculture is faced with the task of not only defining the term, but also coming up with ways to evaluate whether a given approach to tillage or raising animals or marketing is likely sustainable. At the Kerr Center we have devised our own steps to a sustainable agriculture, with goals that address what we see as the key problems of agriculture today.

DOWN TO EARTH: EIGHT STEPS TO THE NEXT GREEN REVOLUTION

In the spring of 1989, I found myself once again in Washington, DC, scheduled this time to speak at the National Conference on Low-Input and Sustainable Agriculture. Attitudes were different from what they had been two years earlier when I had testified to Congress. Sustainable agriculture was no longer like a seed in the ground, waiting to germinate. It had sprouted, raised its young leaves to the sun, and begun to grow. It was now a sapling, with strong roots anchoring it in the soil, and new leaves unfurling each day.

At the Kerr Center, we were growing, too, after a couple of years of rough weather. We hired new staff—people more comfortable with the word *sustainable*—and we were eager to again do what the Kerr Center's predecessor had done successfully for over twenty years: reach out to farmers. We had new demonstration projects and were planning to revive farm consultation teams to advise area farmers about sustainable agriculture practices.

The unanswered question was: Just what does one do to encourage sustainable practices among conventional farmers? I believed firmly in education—that if you gave farmers sensible, honest, science-based information, they would use it, though it would take time. In agriculture everything moves slowly, in sync with the seasons of the year. Change in agriculture is slow partly because experiments with plants and animals take time to show results and partly because change is financially risky. Farmers operate on the slimmest of profit margins. Lenders do not encourage change; they too are wary of unpaid loans. This was especially true in the mid-1980s, in the midst of the most severe farm depression since the 1930s.

The first thing we had to do was get a sufficient grasp of the ideas of sustainable agriculture so that we could employ and demonstrate them on our 4,000-acre ranch/farm and explain them to area farmers and ranchers. One way to do it would be to approach it from the farmer's point of view. What is the farmer most concerned about?

In my speech in Washington that day, I identified a few basic areas of perennial concern to farmers: Pest control, profit, soil conservation. Unless we addressed these areas in ways that the average farmer could understand and adopt, I remarked, we were not going to be successful.

Over the years, I have expanded and contracted these points of concern. The current list has stood the test of time and usage on the ranch and in consultation with farmers. I have thought of these areas as criteria to use when evaluating the sustainability of a project, or as goals to keep in mind when making decisions on the farm. For this book, I have stated them as steps to a sustainable agriculture. They are as follows:

1. Create and conserve healthy soil.
2. Conserve water and protect its quality.
3. Manage organic wastes without pollution.
4. Manage pests with minimal environmental impact.
5. Select livestock and crops adapted to the natural environment.
6. Encourage biodiversity.
7. Conserve energy resources.
8. Increase profitability and reduce risk.

These simple yet comprehensive guidelines would help my staff evaluate the "sustainability" of proposed projects, and would allow farmers to evaluate their own farming practices. Area farmers at the time had little information about sustainable agriculture and little was forthcoming from traditional sources of information. Of course, there were individuals and clubs in the state interested in organic gardening, but we were the first group in Oklahoma to tackle the idea of a sustainable agriculture in any substantial way, committing money and our whole staff to the notion. We were groundbreakers partly because of our status as an independent, nonprofit foundation: we had no bureaucracy to sway or legislators to convince or donors to please.

These goals point to remedies for the three-part indictment in Chapter 1. In the first part of the indictment, industrial agriculture is charged with endangering the essential natural resources of soil, water, and life, thereby jeopardizing the future productivity of the land and the inheritance of our children. If sustainable agriculture is to provide remedies for these wrongs, it must therefore preserve the health of essential natural resources and safeguard the future productivity of the land and the inheritance of our children.

In the second part of the indictment, industrial agriculture stands accused of hooking farmers on fossil fuels, and the fertilizer and pesticides made from them, while downplaying the consequences of overusing such products. A sustainable agriculture would have farmers relying on renewable energy resources and environmentally responsible farming methods.

In the third part of the indictment, industrial agriculture is charged with desolating rural America by bankrupting farmers and ignoring the well-being of rural communities, leaving them open to exploitation. A sustainable agriculture must therefore support the health of rural America by increasing the profitability of, and opportunities for, viable, independent medium-sized and small farms.

These are broad goals. As any farmer (or anyone who has tackled a big job) knows, success often means breaking the job into small pieces. That's what the eight steps do. Farmers, ranchers, and agriculture educators can use these as a guide to sustainable agriculture. Those interested in a healthy environment can look at these and get an understanding of what farmers are facing.

Many of these steps can be used by home gardeners and urban landscapers to promote healthy ecosystems in the city. Consumers can use some of the items on this list when evaluating their choices in the supermarket. For example, if a consumer in Oklahoma in May had to choose between locally produced leaf lettuce and imported-from-California iceberg lettuce, a glance at this list would likely tip the balance toward the locally produced leaf. That is because the California lettuce was probably produced in an irrigated field, using up water resources, while the locally produced lettuce was grown using our abundant spring rains. The local lettuce was shipped only a short distance, while the California lettuce burned up a lot more of our limited supply of fossil fuel (in the form of gas or diesel for trucks) to get here, not to mention the energy used to refrigerate it along the way. Leaf lettuce grows well here in the spring, so it is an adapted crop. Growing such a high-value crop will help farmers make a better profit. And while not always the case, it is likely that the California lettuce was grown on a lettuce megafarm, while the Oklahoma lettuce was probably raised by a smaller producer.

It is an exciting time in agriculture. Although some fear that the term *sustainable agriculture* has become so general that it means little, I agree with those who see it as a catalyst for a new way of thinking about farming. I have high hopes for this new agriculture, the next green revolution.

Chapter 3

Down to Earth: Step 1—Create and Conserve Healthy Soil

> . . . soil is not usually lost in slabs or heaps of magnificent tonnage. It is lost a little at a time over millions of acres by the careless acts of millions of people. It cannot be saved by heroic feats of gigantic technology, but only by millions of small acts and restraints . . .
>
> Wendell Berry[1]

In the spring of 1996, the amount of land in western Oklahoma—1.8 million acres—in a "condition to blow" reached a twenty-year high. A field is in this category when it doesn't have adequate growing cover or residue from a previous crop to protect the soil from eroding with high winds. The immediate cause was a severe drought that had begun the previous fall and lasted throughout the winter. In contrast, the previous year, 1995, had seen a twenty-year low in the amount of land in danger of serious wind erosion damage, with 158,890 acres liable "to blow" in the incessant western wind.[2]

In comparison to the almost two million acres threatened in 1996, the 1995 figure looks great. But is it? This statistic tells us that there are at least 150,000 acres of land in western Oklahoma subject to serious wind erosion damage each year. What do farmers do to combat it? In 1996, they were out on their tractors practicing emergency tillage—plowing up strips to bring clods of dirt to the surface. These strips of clods help break the force of the wind across a field, and, therefore, lessen erosion.

I felt for these farmers. I had spent too many days myself on a tractor, mouth covered with a rag, eyes protected by goggles, doing the same thing when I was growing up. I remember wondering if it was worth it; it seemed more symbolic than effective. Still, we did it—anything to keep our fields from blowing away. Here and there, where tumbleweeds had got hung up in a fence, accidental windbreaks formed. The dirt caught in the weeds would make sand walls above the field. I remember dust storms during the 1950s that buried fences and cars and led my mother to stuff window cracks with rags and hang quilts over the door. One time when we visited my grandparents we took scoop shovels and cleared out five inches of sand that had seeped into their house from one such storm.

They lived in west Texas and, of course, it wasn't and isn't just Oklahoma with this problem. But in Oklahoma blowing soil has a special significance. We are Steinbeck's Okies, "blowed out and tractored out" of our farms during the infamous dust bowl. The blowing fields of 1996 are a reminder of what I saw on a postcard in a store at the mall recently. Labeled "Oklahoma 1935," it pictured a black tidal wave of topsoil on the horizon that threatened to drown all in its path under an ocean of fine dust.

What events brought on the dust bowl? Some were farming land that turned out to be too easily eroded and should never have been converted from grassland to cropland in the first place by the "sodbusters" desperate to make a living.[3] Then came the terrible drought that withered crops and left fields exposed to the wind. Imagine a dust storm carrying 300 million tons of fertile topsoil—the equivalent of a foot of soil stripped from 150,000 acres—about 470 square miles.[4] The dust bowl wasn't just a spectacular natural disaster, worthy of big screen treatment, it was a tragic loss of an essential gift—a gift that makes life on earth possible: topsoil.

NOT JUST DIRT

Topsoil is a constant that we are largely unaware of, a little like the air that we breathe. When a farmer with a plow turns over the soil in a field, or a gardener sticks a shovel in the ground and spades up a mound of dirt, each is encountering topsoil. As its name implies, it is the top layer in a multilayer cake, the bottom layer being bedrock. But it is not just the icing on the cake—it is the layer that makes life on earth possible.

Topsoil consists of mineral particles, organic matter, water, air, and living soil organisms. It is the layer of life—the root zone of plants where the water and nutrients that enable plants to grow are absorbed. Beneath the topsoil is the subsoil that generally has less organic matter and is less penetrable to roots than topsoil, but is a storehouse for minerals.

Topsoil depth varies from place to place. In the Nile River Valley, built by eons of flooding and deposits of sediment, it is tens of feet thick. This luxurious topsoil was the fertile foundation for humanity's progress. Often, however, topsoil is more precious. When settlers first came to the North American prairies, the topsoil depth averaged ten inches, built up over the centuries and held in place by deep grass roots. In the hundred years since the prairie sod was broken by the steel plow and the land was converted to farming, on average about half the topsoil has been eroded away. The creation of the breadbasket of America from the virgin prairies has come at great cost in topsoil.[5]

Topsoil is crucial to agriculture. This first step—creating and conserving healthy soil—is the foundation of a sustainable agriculture. It is closely linked to the next two steps—conserving water and protecting its quality, and managing organic wastes and farm chemicals so they don't pollute. How well agriculture manages soil, water, and organic wastes will determine its future health.

Conserving healthy soil by guarding it against erosion or other forces that would degrade it is the most basic step. This step has as its corollary actively building soil health, because soil used for agricultural purposes today is not as healthy as it could be. It is both less diverse and less active biologically. Without healthy topsoil, the world cannot begin to feed its billions. Although American popular culture discourse in recent years has speculated on the fate of life on earth in case of alien invasion, asteroid bombardment, or rampaging killer viruses, the slow loss of quality soil is more of a threat to life on the planet than any of these scenarios.

This first step—conserving and creating healthy soil—is the foundation of a sustainable agriculture.

Perhaps the relative lack of concern about this problem in the United States has a historic cause. Ever since the first Europeans landed in

North America, the vastness of the continent led them to believe that its natural resources—including soil—were inexhaustible. This belief was held even by those who should have known better. In 1909, the chief of the Bureau of Soils labeled soil "the one indestructible, immutable asset that the nation possesses."[6]

It wasn't indestructible or immutable, then or now. As there are today, back then there were those who ignored the problems in agriculture as well as those who tried to address them. Conservationist and president Theodore Roosevelt recognized the danger and pronounced in his down-to-earth manner: "When the soil is gone, men must go; and the process does not take long."[7] In the U.S. government's Bureau of Soils, one scientist in particular, Hugh Hammond Bennett, spoke out and wrote on the issue and began to study rates of erosion in test plots, including one in Oklahoma. But it took a dramatic event like the dust bowl—when the soil loss was so easy to see—to get people to really pay attention.

Soil erosion—the removal of topsoil by wind or water—is the most dramatic way that topsoil can be degraded. During the terrible drought of the 1930s in the Great Plains, wind picked up soil from withered fields and blew it literally across the country. The severity of the situation caused the U.S. Congress in 1935 to begin considering a bill to create the Soil Conservation Service (SCS), an agency of the federal government that would address the nation's soil erosion problems.

Hugh Bennett was scheduled to testify in favor of the plan on an April morning. Usually concise and to the point, he instead gave detailed information about erosion problems in state after state. As he warmed to his subject, the sky slowly darkened with dust. Bennett, of course, knew what was coming. It was a dust storm, carrying topsoil from prairies 2,000 miles west. While the dust obscured the great white buildings of official Washington, the Congress was persuaded to pass a bill establishing the SCS, the first soil conservation act by any government in history.[8]

The Soil Conservation Service moved quickly to stop or reduce erosion. The government bought some erodible farmland and paid farmers not to farm other marginal acres. Billions of trees were set out in rows at the edges of fields—shelter belts—to break the wind. Blowing fields were seeded with grass. Within two years, the SCS

Types of Soil Erosion

In any given location, various types of erosion may be active and account for considerable soil loss. In other cases, only one or two of these erosion processes exist.

Erosion Caused by Water

Splash erosion occurs when raindrops break the bond between soil particles and move them a short distance.

Sheet erosion takes place when dislodged soil particles are moved by thin sheets of water flowing over the surface.

Rill erosion occurs when the surface flow of water establishes paths called rills, and flowing water readily detaches soil particles from their sides and bottoms.

Ephemeral or *concentrated-flow erosion* occurs when the topography of a landscape is such that rills tend to enlarge and join with others to form channels that are erased by tillage operations but often reform in the same location with each storm.

Gully erosion takes place when concentrated-flow erosion is allowed to continue over time and causes a gully to form. Gully erosion is difficult to control because soil is rapidly removed by water gushing over the "head cut" (uphill end) of the gully, by water scouring the gully's bottom, and by water removing soil material that has slumped from the gully's sidewalls.

Stream bank erosion occurs when the stream flow causes caving and sloughing of streambanks.

Erosion Caused by Wind

Saltation or movement of fine and medium sand-sized soil particles begins when the wind velocity reaches about thirteen miles per hour at one foot above the ground surface. The particles are lifted only a short distance into the air and the spinning action and their forward/downward movement give them extra power to dislodge other soil particles when they hit the ground. Saltation also destroys stable surface crusts, creating a condition more vulnerable to erosion, and the amount of soil moved increases with the width of the field. Saltation accounts for 50 to 80 percent of the total soil movement from wind erosion.

Suspension refers to the process by which very fine soil particles (the fertile organic matter and clay portions) are lifted from the surface by the impact of saltation, carried high into the air, and remain suspended in air for long distances. This "dust" can be blown hundreds of miles and is what most people associate with wind erosion.

Surface creep is the movement of larger (sand-sized) soil particles along the ground surface after being loosened by the impact of saltating particles, but such larger soil grains are too large to be lifted off the surface in most winds. These larger particles move in a rolling motion along the surface and can account for up to 25 percent of the soil moved by wind.

Source: U.S. Department of Commerce, National Technical Information Service, U.S. Economic Research Service, *Soil Erosion and Conservation in the United States: An Overview,* AIB-718 (October 1995).

was working with 50,000 farmers. The work of the SCS, now renamed the Natural Resources Conservation Service (NRCS), has continued to the present day, through a variety of programs. In 1974, conservation practices had reduced soil erosion on the Great Plains by 221 million tons annually.[9]

Unfortunately, despite all these efforts, we still can't declare a victory over soil erosion. Farmers must continually battle it, not just in the old dust bowl of the Great Plains, but across the country and the world. Erosion can be caused by wind or water, and can occur on the country's best farmland—Iowa, Illinois, and Missouri top the list for water erosion.[10] It has been estimated that two bushels of soil are lost from Iowa farmland for every bushel of corn produced.[11] Unlike the dust bowl, which was an acute case of soil loss, erosion today is a chronic problem, exacerbated on occasion by natural disasters such as floods or drought, as in Oklahoma in 1996.

In 1938, the SCS estimated an annual total loss of 3.56 billion tons of soil from cropland. Amazingly enough, in 1982 the loss per year on cropland was just slightly better at 3.1 billion tons. By 1992, the rate had declined by one-third—to 2.1 billion tons.[12] Overall, 67 percent of the soil savings on cropland over the ten-year period came from reductions in erosion on highly erodible land.[13] Much of the credit for the soil savings has been given to the USDA's conservation programs such as the Conservation Reserve Program (CRP), implemented in 1985, which paid farmers to plant grass or trees on highly erodible cropland through ten-year contracts with the USDA.

Although the decrease in erosion is a significant improvement, to the average taxpayer it may seem like progress has been slow, given the more than sixty years of government programs and resources that have been devoted to combating erosion. Unfortunately, the success of government programs such as the CRP may turn out to be as ephemeral as the "ephemeral gullies" that come and go with the rain (and are never included in the erosion statistics). The success of such a program depends on farmer participation which, in turn, seems to depend on the amount of money the government is willing to pay to make it worth the farmer's time and money to participate.[14]

Soil erosion is a natural process that can be worsened by the activities of people. Forty percent of U.S. erosion losses are from nonagricultural activities—logging, construction, off-road vehicles, floods, droughts, and fires. Although agriculturists rightly point the finger at others for squandering a basic natural resource, others rightly point right back at us for being responsible for more than half of the soil erosion in the country.

Unlike the dust bowl, which was an acute case of soil loss, erosion today is a chronic problem. . . .

Soil erosion obviously affects the health of the soil. It seems absurdly self-destructive for farmers to destroy the very natural resource upon which their livelihood depends. And of course many farmers are stewards of the soil or want to be. As then-Oklahoma Farm Bureau President Jack Givens said in an April 1999 press release, "For us in agriculture, every day is Earth Day." He voiced the stance of many who defend agriculture for its environmental record when he asserted that farmers, "every day . . . take pride in protecting and enhancing our natural resources for today and for future generations."[15]

It seems clear, though, that there have been and still are many factors—financial pressures, government programs, and accepted farming practices—that have worked against that stewardship ethic. The changing structure of agriculture, with the loss of so many family-worked farms, has also worked against stewardship of the soil. As farms get very large, there is too much land for the operator and family to manage themselves and really manage well.

In addition, because the effects of erosion are long term, with fertility slowly declining, and because the effects tend to be masked by chemical applications, often farmers do not realize the extent and consequences of erosion on their own land. Or if the quest for short-term profits has become paramount, as on some corporate-owned farms, the long-term health of the soil is not a major consideration.

It is not just erosion, however, that threatens healthy topsoil. A greater threat is the attitude that soil is a lifeless medium for holding plants up and holding fertilizer. In this approach, the farmer adds prescribed amounts of chemical fertilizer, sometimes a blend of nitrate, phosphate, and potash, or nitrate alone, to his field or pasture, sometimes according to a soil test, sometimes not. The fertilizers feed the plant and make the crop. Any natural fertility in the soil is not increased. The idea that the soil itself is an ecosystem—that, if understood and nurtured, would provide a steady flow of nutrients as well as provide other benefits—is not part of the equation.

Now it is undeniably true that crops can be grown chemically—the United States proved this by its spectacular production rates since the widespread adoption of chemical fertilizers. But advocates of sustainable agriculture believe this approach is not sustainable in the long term (as do the organic farmers who, though largely ignored for years, patiently stuck to the belief that soil lies at the center of a healthy agriculture). It seems to me that there are three main reasons why the sustainable folks are right.

First, the agricultural reason—using lots of chemical fertilizer instead of organic fertilizer means the organic matter in the soil is not replenished. Organic matter, as we shall see, is key to the long-term health of the soil. As it breaks down it releases nutrients to plants. Its presence in the soil makes topsoil act like a sponge—holding air, water, and nutrients for plants to use. Farmers ignore organic matter at their peril.

Those who have spent their lives studying the soil, such as Selman A. Waksman, who won the Nobel Prize in 1952 for his discovery of the antibiotic streptomycin (produced by soil organisms), emphasize the importance of organic matter. Waksman has said: "The continuous use of mineral fertilizers on the same soil for many years, without the use of organic manures or growth of sod crops to replace the organic matter lost by clean cultivation, may lead to deterioration of the physical condition of the soil and loss of productivity."[16]

Second, what agriculturists call "off-site effects" of heavy chemical fertilizer use on water and aquatic life have become unacceptable to society. This is an environmental problem that affects large numbers of people.

Third, economics. While fertilizer prices are still relatively low, they are predicted to rise as the amount of raw materials declines, the demand increases, and overseas supplies are subject to disruption.

I agree with sustainable agriculture advocate Marty Strange, who cherishes the values and the neighborliness one finds in family farming communities. But, just as he does, I find fault with the greater agricultural community for too often saying one thing and doing another. There is no system of agriculture that brags more about how it respects the soil yet in reality has respected it so little. When we arrived on the continent, the natural fertility of the soil, especially under the prairie grasslands, was stupendous, so rich crops were grown for years on its natural fertility. This natural fertility has not been replenished. As Strange has written: "We have done more damage to the topsoil of the Midwest in a hundred years of family farming than the communal cultures of Native Americans did in a millennium or the ancient civilizations of the Middle East did in their span. We have been richly endowed and we have squandered it."[17]

A sustainable agriculture must feature farming practices that conserve and create healthy soil and support farmers who put soil first.

FIGURE 3.1. Nutrient Cycle

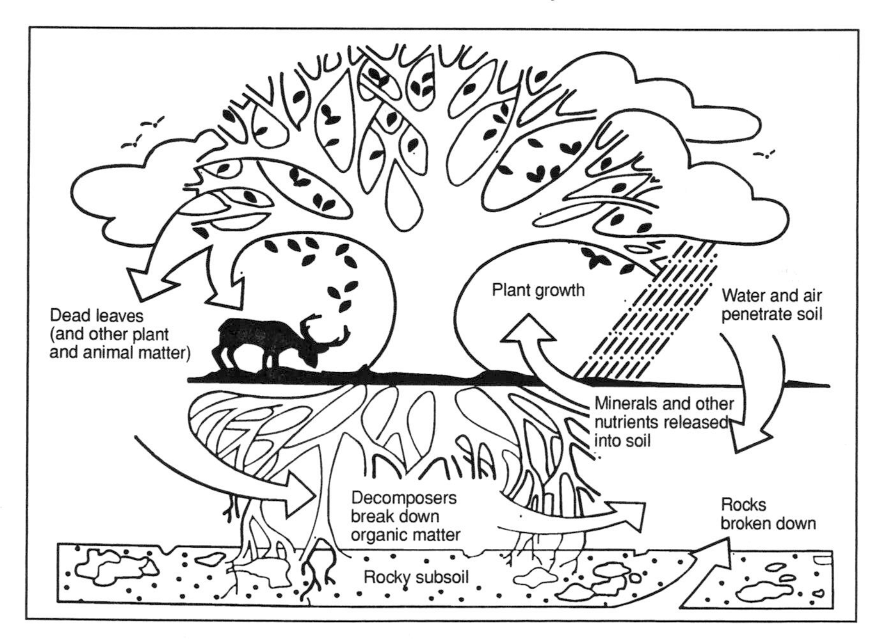

Source: USDA, Soil Quality, NRCS/RCA Issue Brief 5, November 1995.

THE GOOD EARTH

Each spring when the farming year begins in North America, the farmer goes to his fields and pastures and is greeted by the smell of good soil. Like the yeasty smell of baking bread, it is deeply satisfying—a smell of awakening, of beginnings, of fertility, of promise.

Soil is the real staff of life, and every farmer who takes stewardship seriously realizes this basic truth. Along with air, water, and sun, soil makes life possible on earth. It anchors plants and feeds them from its storehouse of nutrients. We, of course, eat plants or the animals that eat plants, and so we are linked to soil in a profound way. Cultures all over the world have long acknowledged the deep connection between soil and humanity. In the Book of Genesis, the first human is fashioned from clay. It is truly the good earth.

When I was a boy growing up in western Oklahoma, our farming year began when we began to work the soil in the spring, preparing the fields for planting cotton. Although winters in Oklahoma are not as severe as those in the more northern prairie states, when spring truly arrives, everyone is happy, particularly because spring teases us for several weeks. First, we get warm spells that cause the fruit tree buds to swell; and then we are slapped with late freezes which cruelly kill all hope of apricots or peaches later in the season.

Back then, my understanding of soil was limited, as was my father's and our neighbors'. The average farmer in those parts knew there was topsoil and subsoil and that you wanted as much topsoil as you could get. (And I don't think the depth of understanding is that much different today.) When it came to soil, we were mostly concerned about "the wind coming up"—in other words, soil erosion. Prices, insects, and weather were always on our minds because those were the unpredictable, uncontrollable factors that made farming so difficult.

Now I raise cattle. Annually, on one of those early warm spring days, we work them—vaccinating the babies and checking the health of the mothers. In my pastures, the fescue and rye put on new growth. Soil, however, is just as important to the cattleman as it is to the farmer. What goes on in the root zone of the grasses and clovers and broadleaved weeds will determine the health of the pasture.

Soil is the real staff of life, and every farmer who takes stewardship seriously realizes this basic truth.

What is soil? Typical soil is made up of four main parts: minerals (45 percent), water (25 percent), air (25 percent), and organic matter (about 1 to 5 percent). Soil is formed when rock decomposes (through weathering and biological action) and when plant and animal life decays. It sounds simple enough, but it isn't. Some say that soil is the most complex ecosystem on the planet.

Although soil is complex and still mysterious, scientists have identified a number of characteristics of healthy or quality soil. These are, in short: good texture and structure, high amounts of organic matter, and active soil life. These characteristics are inextricably linked, with both soil life and structure largely dependent on organic matter. A sustainable agriculture aims to build healthy soil by paying attention to these components.

Probably not enough can be said for organic matter (or o.m., as it is abbreviated on soil tests). Plants, just like humans, need adequate nutrition for optimum growth. Organic matter provides nutrients for plants with the help of the multitudes that live in the soil. Good soil is not dead dirt; it's very much alive. Moles, gophers, prairie dogs, spiders, dung beetles, worms, centipedes, millipedes, snails, slugs, bacteria, and fungi are just a few of the animals both seen and unseen living in soil. Many help build soil and battle soilborne plant diseases. The microorganisms in just one spoonful of soil outnumber the people on earth. The numbers are astronomical—seventy-one billion bacteria to the ounce in a fertile agricultural soil,[18] or, if you prefer, 800 quadrillion to the acre.[19] Microorganisms are essential to the chain of life, breaking down organic wastes and releasing nutrients to plants. Soil health is a direct function of the biological activity in the soil. It is this biological energy that sustainable agriculturists say can be enhanced and used to fertilize crops without additions of chemical fertilizers.

This cycle of birth, growth, reproduction, death, and decay is a powerful one. A case in point: a crop, say corn, is planted. It grows and reproduces, and its seed, the kernels on the ears of corn, are harvested. The parts of the plant not used, the leaves, stalks, and roots, are left in the field. As the remains of the plant come into contact with the soil, they begin to decay, eaten by bacteria and fungi.

The microorganisms in just one spoonful of soil outnumber the people on earth.

So begins the process of transforming complex organic compounds of living matter into simpler compounds that plants can use for food. The corn residue becomes organic matter. From the animal and vegetable proteins in this organic matter, bacteria form ammonia. Other bacteria transform the ammonia into nitrite, and then still other bacteria turn nitrite into nitrate, a type of nitrogen that is used by plants. When the next seed sprouts in the old corn field, its roots absorb this form of nitrogen, an essential nutrient. The cycle begins again. In agriculture, this process is known as nutrient cycling.

A substance essential to nutrient cycling is humus. If you picked up a handful of rich soil, you would be holding it in your hand. It is organic matter in the soil that has reached an advanced stage of decomposition. It is a dark color, and high in nitrogen. The Romans recognized its importance: Humus, in Latin, means soil. Humus is the heart of healthy soil. Howard called the positive effect of humus on crops, "nothing short of profound."[20]

Besides adequate nutrition, plants need adequate water to grow. The other key feature of healthy soil is good soil structure: soil granules and the right amount of pore space. Pore space allows nutrients, water, and air to move easily through the soil and allows roots to develop extensive networks for absorbing water and nutrients. Just as bread is dense and heavy when it lacks air bubbles, so soil without air is heavy and dense. The big tractors used on contemporary farms can cause poor soil structure because their weight tends to compact the soil, squeezing out those air spaces and water spaces. Adequate organic matter lightens the soil and creates pore space. Organic matter is also like a sponge, absorbing water. The amount of water that can be stored in the soil, available to plants as they need it is increased. Because of this, adequate o.m. also prevents runoff topsoil erosion.

If the soil is a world of its own, its unsung heroes are the earthworms. Their tunnels aerate the soil, improving its structure, and they eat organic matter and leave nitrogen-rich castings as food for plants. Great minds have found much to admire in the lowly earthworm. As Charles Darwin wrote in 1881, "The plow is one of the most ancient and most valuable of men's inventions but long before he existed the land was in fact regularly plowed, and still continues to be thus plowed by earthworms. It may be doubted whether there are many other animals which have played so important a part in the history of the world, as have these lowly organized creatures."[21]

These natural connections between organic matter and soil life unlock nutrients and make them available to crops. Heavy use of synthetic fertilizers (and pesticides) can limit the ability of soil organisms to process wastes. In contrast, recent research has found that adding organic matter to soil causes much greater biological activity than the use of a synthetic nitrogen fertilizer. (In the study, equivalent amounts of nitrogen were added.) The researchers found that soils with a long-term history of organic matter additions had higher microbial activity, which correlated with the amount of carbon added. They also found that adding new organic matter to soil caused a large increase in microbial activity, whether the soil had a history of such additions or had just been treated with synthetic fertilizers. Adding new synthetic nitrogen alone had only a limited stimulating effect.[22]

Another recent study has confirmed another benefit of healthy soil rich in organic matter. An Ohio researcher found that there was a general relationship between plant nutrient levels and insect damage (in this case, the corn borer). But pest resistance seemed mainly to be affected by the balance of nutrients in the crops—ratios of particular nutrients, rather than absolute levels. Organically managed soils promoted a desirable nutrient balance, possibly by dampening fluctuations in water and nutrient availability to plants.[23]

Adding organic matter increases fungi that "eat" destructive nematodes that live in the soil. Indeed, it has been observed in other parts of the world that crops grown in traditional ways on soils high in fertility and organic matter in general suffer less from pests.[24]

In the long run, good soil is not made by spreading chemical fertilizer on a field year after year.

Chemical fertilizers used heavily cause the soil pH to become more acidic, which makes it more difficult for plants to utilize nutrients. However, chemical fertilizers have a number of positives, too: they are easy to buy and apply, their price is low, their elements are known, they are accessible to plants, and they act quickly. I believe they have a place in agriculture at current prices. Unfortunately, they are most often used to correct problems caused by poor agricultural practices such as overgrazing or monoculture. Furthermore, reliance on synthetic fertilizers also tends to exclude farming methods that would increase organic matter.

In the long run, good soil is not made by spreading chemical fertilizer on a field year after year. Making good soil requires a more complex recipe: time, the right ingredients, and the right system to reestablish the natural cycles of fertility interrupted by industrial farming practices.

The Gift of Good Land

Although many of those prominent in Oklahoma agriculture today have not been particularly receptive to sustainable agriculture, some in an earlier generation were pioneers in the field. Joseph A. Cocannouer, who once lived about fifty miles from the Kerr Center, wrote eloquently about soil in his 1954 book *Farming with Nature:*

> A chemical analysis of soil, infinitely valuable though it is, still does not give anything resembling a complete picture of the soil's worth, for chemistry deals primarily with the dead, not the living. Active biology, though, helps greatly to fill out a reliable chart of information. A naturally rich soil is so alive it seems to move when you hold it in your hand. Then there are the rich aromas produced by organic decay and microbes and molds. The latter are unmistakably indicative, whether emerging from a healthy soil itself or from the healthy growth that springs from such a soil. The feel of soil can also be a reliable gauge of the soil's richness. One does not forget the touch of highly fertile earth which is dark, crumbly, non-caking when wet. Working in a soil with all of these natural attributes brings one very close to the true potentials for plant growth.[25]

Another seminal thinker on these things was British agriculturist Sir Albert Howard. In his 1940 book *An Agricultural Testament,* Howard neatly summarized what he called "Nature's farming," or how a natural ecosystem such as a forest maintains itself. The following passage has served as inspiration to those who want to practice sustainable agriculture since it first appeared over a half-century ago. Note the great emphasis Howard put on conserving and building soil in this passage:

> The main characteristic of Nature's farming can therefore be summed up in a few words. Mother earth never attempts to farm without live stock; she always raises mixed crops; great pains are taken to preserve the soil and to prevent erosion; the mixed vegetable and animal wastes are converted into humus; there is no waste; the processes of growth and the processes of decay balance one another; ample provision is made to maintain large reserves of fertility; the greatest care is taken to store the rainfall; both plants and animals are left to protect themselves against disease.[26]

The farmers of my youth were not schooled in Nature's farming, in what organic matter could do for the soil. But I believe on some level we knew, because there's an intuitive good feeling you get whenever you turn something green into the soil. My spine tingled when I plowed clover under; I knew it was good but I didn't really know why. And I didn't discover why during my college days when soils class discussion centered on how to add nutrients to soil through chemical fertilizers. I don't recall "soil health" ever being mentioned, much less stressed. I began to understand its great importance when I began reading *Organic Farming and Gardening*. And then, although I was convinced that their ideas about healthy soil were valid, I didn't believe they could be taken from a garden and applied on a larger scale to farms.

Now I know they can. Farmers all over the country are discovering what Kenneth Repogle, a northeastern Oklahoma soybean farmer, has discovered. When we visit his 2,000-acre soybean operation, he insists that we smell his soil. Repogle has changed his farming practices in the last few years, adopting ridge tillage—a new approach that minimizes tilling the soil and leaves a mulch on top to be converted into organic matter.

Repogle says his soil is rich with earthworms again. "I haven't seen worms in the field since I was a kid, tagging along behind Dad's 8N Ford and two-bottom plow," he said in an interview in the *Oklahoma Farmer-Stockman*. He has taken to carrying a hoe or shovel around with him just to keep tabs on the improvements in his soil.[27]

He has noticed too how much better his soil absorbs and holds water: When it rains heavily the water doesn't stand in the field like it used to;

and when it is dry, his soil stays moister than his neighbor's, who uses conventional tillage methods.

Ridge tillage is just one of a number of strategies that sustainable agriculture suggests will upgrade the nation's soils—strategies that had largely been out of favor with industrial agriculture. These methods decrease or stop erosion and increase the health and long-term fertility of the soil. At the Kerr Center, in the past fifteen years we have tried to be the best stewards we can be of our little patch of the good earth by conserving and creating healthy soil in our pastures and more intensively, on our horticulture farm.

When I was in high school, I gave a speech titled "Our Soil—Our Freedom" for a Future Farmers of America competition. In it, I made the case for soil conservation. I began by quoting Exodus 3:5: "The place upon which thou standeth is holy ground." Along with the many practical arguments for creating and conserving healthy soil, I believe that land is sacred and that humanity is charged by a higher power to be good stewards of this great gift of good land.

Cover Your Soil

Plowing disturbs the soil. What are the synonyms for disturb? Try distress, disrupt, disorder, agitate, interrupt, unsettle. . . . None is positive. In recent years, American farmers have been discovering the advantages of disturbing the soil as little as possible, and by doing so are allowing the soil to resume the natural processes that had been interrupted. On cropland, this usually means reducing tillage.

Tilling the soil in a conventional way means plowing with a moldboard plow that turns the top eight inches of soil completely over, burying crop residue, and leaving the soil loose and prone to erosion, as my father knew. Reducing tillage, in contrast, means abandoning the moldboard plow and leaving crop residues on top of the ground to cover and protect it from wind and rain. These methods of "conservation tillage"—where at least 30 percent of the field is left covered after harvest—are gaining in popularity.

Conservation tillage systems leave substantial amounts of crop residue evenly distributed over the soil surface which reduce wind erosion and the kinetic energy impact of rainfall, increase water infiltration and moisture retention, and reduce surface sediment and water runoff. (Some no-till systems leave the field 60 percent or more covered.) There are a number of ways to achieve this—some of these practices are no-till, mulch-till, and ridge-till. Each prepares a seedbed using implements other than the moldboard plow. The number of times the soil is "worked" is reduced. Erosion is cut significantly.

In 1997, conservation tillage systems were used on 109.8 million acres, 37 percent of the nation's cropland. Conventional or "clean" tillage (using a plow) in con-

(continued)

(continued)

trast, was still being used on 107.6 million acres. (It leaves less than 15 percent of the ground covered.) This was a milestone. For the first time ever, farmers were planting more crop acres using conservation tillage than conventional.[28] In response to the trend, implement manufacturers are turning out new planters and grain drills that can help the farmer, as one ad put it, "breeze though tall heavy stubble." The USDA has set a goal of conservation tillage used on 50 percent of total crop acreage by 2002.

No-till farmer Ron Jacques, who lives near Hutchinson, Kansas, lauds the advantages of using no-till. By adopting conservation tillage techniques, Jacques and his neighbors significantly reduced the sediment runoff into a nearby lake by 3,500 truckloads of soil per year. No-till, he says, increases organic matter that allows the soil to store and transmit water better.[29] Earthworms like less tillage too—according to the University of Missouri, after one particularly cold winter researchers dug up only five worms under each square meter of tilled plots. They counted 144 worms per square meter in no-till plots. Even after a mild winter, there were three times more worms in the no-till plot.[30]

Abandoning the plow has had another more global impact. According to Floyd Horn, administrator of the USDA's Agricultural Research Service, the switch from the plow to conservation tillage has turned American farm soils from net carbon dioxide producers to net accumulators of carbon in the form of soil organic matter. This makes American "soils more productive and part of the potential global warming solution, rather than part of the problem," he said. He cites research done by Raymond Allmaras, an ARS soil scientist. "The soil is storing more carbon that otherwise might be in the atmosphere as carbon dioxide," Allmaras said. "The plow lifts and inverts an 8- to 12-inch slice of soil and also buries stubble and other unharvested crop residue that was once on or near the surface. That places the residue deep in the plow layer where different microbes live. These microbes convert the residue into a form of carbon that readily converts to CO_2 which can escape to the atmosphere," he said. As farmers put aside the plow, they leave more residue on the soil or within a depth of four inches. "The residue readily decays to valuable organic matter, a more stable carbon compound and a key component of the black, fertile prairie soil originally broken open by the plow."[31]

Although the positive effects of these practices cannot be ignored, they often require herbicides as a substitute for tillage to kill weeds. It is questionable whether relying so heavily on herbicides is sustainable—they can increase water pollution, and they are expensive. Herbicides, because they are made from petroleum products, may not be affordable in the long run. There is also the danger of herbicides damaging beneficial soil organisms.[32]

Results of a recent SARE research and education grant in Ohio showed that there are ways to reduce the negative effects of herbicide use in low-till systems. They tried a number of approaches such as switching to safer products (chemicals with less residual effects), applying in a band next to the crop rather than spraying the entire field, and using high-residue cultivation. Using these approaches with reduced rates of application allowed herbicide reductions of 85 to 95 percent. They also used a small-grain cover crop to help suppress weeds in soybeans.[33]

No-till has been more successfully used in the north, where there are fewer diseases. In Oklahoma, a recent reduction in conservation tillage acreage has been attributed to the conducive environment that crop residue provides for

(continued)

(continued)

wheat scab, a disease that attacks wheat fields and leaves behind shriveled, discolored grain.[34] Approaches other than no-till, while avoiding the moldboard plow, can achieve moderate amounts of residue coverage with perhaps less herbicide use. For example, one can also achieve 30 percent residue coverage by one fall chiseling (with straight shanks), a shallow disking in the spring, field cultivation, and planting.[35]

HEALING THE SOIL

Sapped by overfarming, particularly cotton farming, and thinned by erosion, many of the soils on the ranch and in the area had declined since intensive agriculture began here at the end of the nineteenth century. Earlier agriculturists in the area lived more lightly on the land. They were the Spiro Mound people who built a grand city in the Arkansas River Valley north of the ranch around 1000 A.D. They were part of the Mound Builders civilization, which thrived along the Mississippi and its tributaries. They built their advanced civilization on corn, the miracle grain of the New World. After the Mound Builder civilization crumbled and dispersed, Caddoan-speaking tribes lived in this area. They also had little negative impact on the land, growing corn in the river valleys and hunting for plentiful game in the vast pine, oak, and hickory forests of the uplands and in the hardwoods along the rivers.

In the 1830s, the Choctaw Indians were moved to southeast Oklahoma by the United States government from their home in Mississippi. For the most part, they too practiced small-scale subsistence farming. Ostensibly, this was Indian Territory, but after the Civil War, non-Indians of all descriptions—outlaws such as Belle Starr, itinerant laborers, and especially tenant farmers—began moving into the area. The Choctaws began to lose control of the land after the federal government forced them to break up their common holdings into individual allotments in the late 1800s.

At the beginning of the twentieth century, there was heavy Euro-American immigration into the area, and the population increased dramatically. Much of the land passed into non-Indian ownership. Indian Territory land wasn't free to homesteaders as it was farther west to those who made the famous land runs into Oklahoma Territory. But both in the eastern and western part of the state, these new settlers largely viewed Oklahoma as the Promised Land, their last chance, it seemed, to prosper on their own land.

Sadly, the promise was short lived. Before long, over half the farms in the state were small holdings operated by tenants.[36] Crop prices fluctuated wildly, and there was too much pressure on the land.[37] According to the Le Flore County Conservation District, "deforestation, overgrazing, and erosion from tillage led eastern Oklahoma into a tragic period," namely the Great Depression of the 1930s. Thousands of acres of topsoil were lost to sheet erosion, and the rains carved great gullies into the landscape.[38] Hopes of self-sufficiency were dashed, and, for many, California became the new Promised Land.

Today most of the agriculture in Le Flore County, where the Kerr Center is located, centers on livestock. Beef cattle ranches and chicken farms have replaced cotton and other row crops. This would appear, from a soil conservation standpoint, to be a step in the right direction. There is generally less erosion from pastures than from cropland, simply because the soil has a living cover to hold it in place. In fact, data from the NRCS show that converting erosion-prone land to pasture is a good way to minimize soil loss, as pastures have an average soil loss of about one ton to the acre, as compared to soil loss from cropland that averaged 5.6 tons to the acre in 1992.[39]

Soil erosion is more severe on rangeland. A little over half of the land in the United States, 1.2 billion acres, with 230 million acres of that in Alaska, is classified as rangeland.[40] Rangeland is common in the arid, western United States in states such as Wyoming, Texas, and Montana. These grasslands have erosion rates similar to those of cropland. "Accelerated soil erosion" threatens at least one-fifth of all rangeland acres.[41]

Saving Soil with Rotational Grazing

On grazing lands too there are a number of strategies to stop erosion. One is controlled or rotational grazing (also known as cell grazing, intensive grazing, or management-intensive grazing). The vegetation on grazing lands is a renewable natural resource and should be managed as such. Although rotational grazing can provide many benefits—

including improved water quality, biodiversity, and, for ranchers, more profits, it also positively impacts the soil.

Most cattle in the United States are allowed to graze continuously. In that strategy, cattle can wander where they will and eat what they want. In rotational grazing, a pasture or range is divided into smaller units called paddocks. Livestock graze in one paddock until the rancher has determined the forage in the paddock (sometimes called the cell) has been grazed enough. The rancher then moves the livestock into the next paddock and so on through the paddocks. By doing this, the rancher controls where the cattle graze and how much.

After a paddock has been grazed, it is allowed a little R and R, rest and regrowth. The grasses and other plants in the paddock are allowed to grow undisturbed by livestock until the rancher sees that it is ready to be grazed again.

How does this improve soil quality? Rotational grazing can decrease soil erosion by improving plant cover. Under a continuous grazing system, cattle tend to go back to and eat the plants they like best. Eventually they can overgraze those plants—keep eating them down so short that their nutritional reserves are depleted and they can't recover without heavy fertilizing. The plants die or don't grow as large, leaving patches of bare ground which can erode. Rotational grazing prevents this by forcing the cattle to eat what is available in the paddock, whether it is their favorite meal or not. The pasture is grazed more uniformly and the cattle are moved out of the paddock before the forage is grazed too short.

By having a water source in each paddock, rotational grazing also stops soil erosion by keeping cattle from continually congregating at one or two water sources, such as ponds or streams, where their hooves can erode the banks. Ranchers can in general more effectively control the movement of their cattle, keeping them away from fragile stream banks. Rotational grazing also cuts erosion from cattle trails—those two-hoof-wide lanes of bare dirt cows trample as they crisscross grazing lands.

Rotational grazing also helps the rancher utilize more fully a natural, on-farm fertilizer—livestock manure. The manure is distributed much more evenly across the pasture in a rotational grazing system. This allows nutrients to be cycled more uniformly across the land, rather than being too concentrated in one area, while other areas receive little.

Rotational grazing also helps the rancher utilize more fully a natural, on-farm fertilizer—livestock manure.

Rotational grazing was first tried on the Kerr Ranch in the mid-1960s. The Kerr Foundation was progressive and demonstrated what were thought to be the best ideas in ranch management. This tradition of demonstrating the latest ideas goes back to Senator Kerr himself.

In the 1950s, the senator had established the ranch with quality beef production in mind. He raised prize registered Angus cattle. The management was progressive for the time, the emphasis being on maintaining fertility (through the use of chemical fertilizers), improving cattle through bringing in superior bloodlines, and raising cattle that had highly marbled beef—the kind perfect for the backyard barbecue. To say that Senator Kerr was "into" his cattle might be an understatement—his pride in the ranch was reflected in the bull-head faucets in his bathroom. Big steaks on the barbecue went hand in hand with gas guzzlers and cheap fertilizer.

After the senator died, and the foundation was established with an agricultural division that maintained the ranch; the purpose continued to be demonstrating (though in a more formal way) the best management ideas. Rotational grazing was one of them. When I came on board in 1972, we differed from others doing rotational grazing in that we moved the cattle according to the condition of the forage, not by the then-common notion of a prescribed formula of so many days on and so many days off, such as seven days in/seven days out.

However, on the ranch in those days we had a high stocking rate (number of cows per acre) and fertilized heavily to keep the grass growing. Because of the high stocking rate, we were always on the edge in terms of risk. Adverse growing conditions, such as drought (unfortunately, Oklahoma is drought prone) could upset the balance, so we were tied to heavy usage of synthetic fertilizers (as well as weed sprays). Though the system was more progressive than a continuous grazing scheme, it was not sustainable.

Today, rotational grazing is the centerpiece of our management of the 1,500-acre ranch at the Kerr Center (there are 1,000 acres of open grassland; the rest is woodlands and four small lakes). It differs from what we used to do in that it is now a low-input system. No commercial fertilizer or lime (used to raise the pH of the soil) has been spread on the area that is rotationally grazed since 1988. In fact, we are attempting to maintain fertility through rotational grazing—by grazing forages to the height at which they are not stressed and by taking full advantage of our herd's manure. The amounts returned to the pasture to be recycled vary according to the age and condition of the animal and the nature of the feed, but the return can be significant. A mature cow (not pregnant) can excrete 75 percent of the nitrogen and 85 to 90 percent of the phosphorus and potassium she takes in, which goes back to the soil.[42] This is another form of nutrient cycling.

Our goal has been to study the changes in soil fertility, forage productivity, and plant diversity. We learned that rotational grazing is an excellent tool to manage livestock and forage. We have tested the soil every few years to see how the health of the soil is faring. The result: no significant changes in soil fertility and organic matter since 1985, despite not applying any synthetic fertilizers. The nutrient levels are generally adequate. We believe we have generally enhanced biological activity in our soil and unsaddled ourselves of input expenses. Forage productivity has remained high.

Another demonstration project on the ranch is the horticulture farm. It is twenty acres of grass, woods along a creek, and nine acres of cultivated land divided into one- to two-acre blocks, about 1.5 miles from the Poteau River. From 1993 to 1997, we grew tomatoes, sweet potatoes, onions, strawberries, and blackberries organically. (We were certified organic by the state of Oklahoma in 1995.)

In April, the view from the horticulture farm is beautiful—along the creeks and roadsides, the oak trees are unfurling their catkins. Not far away is the greening hillside of Cavanal Hill—known locally as the highest hill in the world, by virtue of its altitude of 2,999 feet, leaving it one foot shy of mountain status. Although the surroundings are ideal, the soil, to begin with, was not.

The soils are silt loams. Some plots are very poorly drained. Levels of organic matter, essential to natural soil fertility, were dismally low in 1993 (1 to 2 percent; organic growers like 4 percent). Whatever the drawbacks, however, the horticulture farm soils are typical of soils in the area, and therefore perfect as a demonstration of what can and can't be done to make such soils healthier. Horticulture farm manager Alan Ware and his staff formulated a system using rotations of cover crops and cash crops and additions of compost to rejuvenate the farm soils.

These are tried and true techniques long employed by organic farmers, and crop rotations, in particular, were once widely used by all farmers to maintain soil fertility. This changed after World War II when farmers began to rely heavily on chemical fertilizers to provide the major nutrients to their crops. Traditional ways of replenishing nutrients in a field fell out of favor.

Compost and Crop Rotation to the Rescue

Crop rotation is an ancient way of restoring fertility to a field. It works because different crops add or absorb different amounts and kinds of nutrients from the soil. The Romans had a system of planting grains, then legumes (bean family). Legumes have the remarkable ability to use the essential nutrient nitrogen from the air.

Using cheap chemical fertilizers makes it possible to skip rotations and grow cash crops, such as corn, on the same field each year—a system known as monoculture. This intense cultivation has resulted in vast increases in production of basic commodities. For the American farmer, using chemical fertilizers makes sense. Although fertilizer prices went up during the 1970s oil embargo, they are still affordable today. (Prices for fertilizers in real dollars in 1992 were actually 25 percent lower than in 1975.) However, prices in the future may well be much higher as the price of oil goes up, as it is projected to continue to do. Also, supplies of potash and phosphorus, the other two of the three common nutrients applied to soil, are projected to be tight in the future.[43] There are other costs to using fertilizers that are generally not figured in—water pollution from the chemicals themselves and negative changes in soil chemistry/processes caused by fertilizers are now recognized as two major problems caused by overuse.

On the other hand, many benefits come from using the older, tried and true techniques. Rotations can also cut soil erosion. In Iowa, using three-year rotations of corn, wheat, and clover, there was, on average,

2.7 short tons per acre lost (NRCS's "tolerable limit" is five tons). Growing wheat continuously lost four times as much soil; growing corn, seven times as much.[44]

Cover crops, as part of a rotation, offer many soil-building benefits. Cover crops are crops grown not for harvest, but either to enrich the soil (then called green manures), to protect it from erosion, or both. A long list of different plants can be used this way: both legumes such as alfalfa and clover, and nonlegumes such as buckwheat or rye. They can be annuals, biennials, or perennials, and can be planted at almost any time of the year.

Cover crops, as part of a rotation, offer many soil-building benefits.

If nature abhors a vacuum, it also abhors a bare piece of ground. Soil is meant to have something growing on it, and growing cover crops puts farmland in sync with the natural principle. We didn't think of it quite that way back in Cold Springs when we grew Austrian winter peas as a cover crop for a few years when the government had a program to pay for such practices. We knew the peas kept the soil from eroding, but our finances were such that when the government stopped the program, we stopped planting cover crops.

After April 15 each spring at the horticulture farm, a summer cover crop was planted, and each fall a winter cover was put in. The cover crops varied widely—some, such as buckwheat, are low-growing and succulent; others, such as crotellaria, in the hemp family, are towering and fibrous. In general, cover crops grow until first bloom when they are at peak mass; then they are tilled into the soil.

This "biomass" increases both organic matter and nutrients in the soil. Legumes used as cover crops can supply nitrogen to the soil, replacing much if not all of what is needed to grow cash grains. How they do this is through a process of nitrogen fixation. *Rhizobia* bacteria live symbiotically in the root nodules of legumes. These bacteria "fix" or convert nitrogen from the air into a form usable by the plant. Any nitrogen the plant doesn't use for its own growth is returned to the soil to be used by the next crop in the rotation. In the northeastern United States, farmers planting winter vetch between their corn rows in late August to be plowed under in the spring can add 150 pounds of nitrogen per acre, enough to supply the needs of the next corn crop.[45]

At the horticulture farm, leguminous cover crops included hairy vetch, clovers, and purple-hulled peas (also known as cow peas). Grasses, too, were incorporated as green manures to increase organic matter and suppress weeds; we used annual rye for winter and sorghum Sudan grass for summer.

If deep rooted, cover crops can bring up nutrients from deep in the soil and make them available to more shallow-rooted plants. Because cover crops increase organic matter, they improve soil texture. Cover crops hold the soil in place and break the impact of raindrops in hard rains. Erosion can be reduced to near zero.

Compost is another soil builder. Horticulture farm manager Alan Ware has made compost from mixing chicken litter from local broiler houses with hardwood sawdust from a local mill, and spread it on various plots in the fall. (The mixture was usually one-third sawdust to two-thirds litter—at rates of two to six tons per acre, depending on soil tests and the litter's nutrient analysis.) It too added organic matter to the soil.

On-farm nutrients, such as animal manures, livestock bedding, straw, old hay, and other wastes, can also be composted. Good, rich compost is an excellent soil amendment, especially desirable to those choosing to decrease their use of synthetic, petrochemical fertilizers. It improves soil tilth (structure), adds nutrients, improves soil moisture holding capacity (decreasing runoff of valuable nutrients), and increases organic matter. It contains not only the major nutrients but also trace elements needed for healthy plants. Compost helps sandy soils hold moisture and improves aeration in heavy clay soils. It minimizes plant yield reductions during times of drought and reduces waterlogging of plant roots during times of heavy rain.

Cover crops and compost improved the growing condition immensely, says Ware. Levels of organic matter have doubled in four years. The pH has increased. This is important because nutrients become more available to plants with a pH closer to neutral. The soil structure has improved, becoming more crumbly and mellow. Without using synthetic fertilizers, the yields of cash crops, such as strawberries, have been decent.

Farmers all over the country who are making the transition to sustainable agriculture are using compost and manures to build their soil. (See the box "Putting Life Back into the Soil" for a firsthand account.)

It is deeply satisfying to take land that has been crippled by misuse and heal it. To Oklahomans, because of our history, it is doubly meaningful. Though this may seem laughable to those in urban areas, I am proud that our state has included an official "state soil" among its emblems: Port silt loam, a rich soil found in the river bottoms of our state. But making sure that good soil is more than symbolic means teaching farmers the best techniques for building soil. We're hoping that the lessons we've learned on the ranch and horticulture farm will help farmers in some small way conserve and create healthy soil. If it can be done here, it can be done anywhere.

Checklist for Farmers:
How to Conserve and Create Healthy Soil

1. Stop soil erosion by planting on the contour, terracing, strip cropping, and repairing gullies.
2. Add organic matter (with "green manure" cover crops, compost, manures, crop residues, and organic fertilizers).
3. Use conservation tillage.
4. Plant windbreaks.
5. Rotate cash crops with legumes, hay crops, or pasture.
6. Employ rotational grazing.
7. Reduce compaction of soil by not working wet soil and cutting the number of trips across fields with heavy farm equipment.

Putting Life Back into the Soil

As an organic farmer, folks ask why there is a three-year transition period required before land can be certified. If chemicals break down within a year, one year without chemicals should be enough to certify land as "organic." I remind them that organic farming is more than mere chemical-free farming. It involves working with the soil and encouraging the life within it. I then tell them a story of the farm I purchased about 10 years ago.

This was the land my dad tried to buy twenty-four years before. It is the flat land, surrounding the hills of my dad's farm, that I watched with controlled envy as others farmed it. So, it was a great day when I finally owned it and could farm it myself. I was really excited as I began to till that first spring. I watched through my rear-view mirror as the disc leveled the ground and covered last year's crop residue.

I finished the new field and moved to the hilly field adjacent to it, dropped the disc and immediately choked the tractor. I was stunned. What happened? I looked for a hidden fence post or tree root that may have caught the disc, nothing. I climbed out to investigate. The soil from the old field was crumbly with tiny roots and insect holes. The lump from the new field was just that, a lump. There

(continued)

(continued)

were no tiny roots, insect holes and no soil crumbles. I was disappointed. This new farm that everyone said, "really laid nice" was only a dirt farm.

I looked further at the lumps and found evidence of crop residue from previous crops. I remembered seeing standing water in the tractor tracks of this farm after a moderate rain. I decided that I needed to work at putting soil life back in this soil. This farm had to do more than just lay nice if it was to pay for itself. I adjusted the depth control on the disc and continued tilling the old farm, making plans for next year.

My thought was to incorporate back into the soil the bacteria needed to break down crop residue. The next spring I applied all the compost I had to the new farm inoculating the soil with a "shot of soil life" and hoped it would grow. I abandoned the corn/soybean rotation for corn/oats with sweet clover. I fertilized with compost and manure instead of anhydrous and chemical fertilizer.

After 3-4 years, the tillage tools now pull the same in all fields. I no longer find prehistoric residue in the soil. Water no longer stands in the tractor tracks after a rain. The building of this soil life—that is what organic farmers talk about when they say farming organically is more than doing without chemicals. It is working with the natural cycles that makes customers willing to pay more for their food. It is the healthy soils that produce the healthy food.

Source: Martin Kleinschmit, Center for Rural Affairs, *Beginning Farmer Newsletter,* June 1999, p. 4.

Chapter 4

Clear, Clean Water: Step 2—Conserve Water and Protect Its Quality and Step 3—Manage Organic Wastes to Avoid Pollution

> We are driving through the earth's resources at a rate comparable to a man's driving an automobile a hundred and twenty-eight miles per hour . . . and we are accelerating.
>
> John McPhee[1]

After big springtime rains in Kiowa County, Oklahoma, my mother would send me out with my little red wagon to dig dirt out of the grader ditch along the side of the road. I piled the soil into my wagon and hauled it back to the garden behind the house.

The family garden was at the base of one of the odd, solitary granite hills that marked the mostly flat land. Although these hills seemed disconnected to the Wichita Mountains further east in Comanche County, they were actually outposts of this ancient range. I didn't know then that the Wichitas had once been as high as the Rockies, but over the millennia had been eroded down to their granite hearts.

Similarly, I took the soil in the grader ditch at face value and did not realize that it had been washed from my mother's garden and my family's fields. I didn't know that my mother's garden soil was poor and that the yields from our fields were declining. I just knew that after a spring rain I liked to watch the water, brown with suspended soil, roar down the ditch into Otter Creek. I liked playing in the deep dirt left by the floodwater in the ditch—my friends and I liked to pretend we were digging for oil, another Oklahoma reality we took for granted. I liked eating the wild greens called lamb's quarters we gathered each spring, and although I knew that we always found the tallest and healthiest of these plants in the deep dirt along the ditch, I didn't know why.

It wasn't until I was older that I understood that our farm's soil was slowly declining in quality as the richest part, the topsoil, was carried away in those big spring rains that polluted the water. Kiowa County receives only between twenty-six and twenty-seven inches of rain on average every year, but almost eleven inches of it comes in April, May, and June. A single spring storm might drop three inches of water. Often hail and sometimes tornadoes accompany such storms, which we rode out in the storm cellar. The cellar was no more than a hole about six feet deep in the hillside, with the walls boarded up to keep the soil from caving in. The top half of the cellar was covered with a mound of dirt; on top of the dirt sat large rocks.

"Going to the cellar" was a common experience in rural Oklahoma in those days, but it was not something we looked forward to. The place was dark and moist and home to snakes. I learned a lot about silence in that cellar; and I learned to fear "the look." Sitting in the cellar with my parents and brother, I remember feeling helpless as we waited for the storm coming from the west. The tension rose as the winds increased and we sat huddled, waiting for Mother Nature to give us either blessed rain or cursed hail. We always tried to guess the intensity of the storm by venturing out to peek at the color of the sky (green being the most ominous), and then speculating on what it would do until the wind wouldn't allow it. Then we would hear either the welcome din of raindrops hitting the sheet metal cellar door or the dreaded *ping, ping, ping* of the first small hailstones, a precursor to the pelting of the larger hail that was almost sure to come. As the hail rained down on the cellar door and the noise increased, talk ceased. Everyone in the family shared the same intense emotions. The expression on my parents' faces said it all; this was "the look" I dreaded. As we waited in the dim light, I visualized the scene in the field: cotton plants stripped of all leaves and the stems beat to the ground.

After the storm let up, we would walk first to the field, not to the house to look for broken windows or chipped paint or a leaking roof. We could easily replace a window, but not a crop. That walk with the flashlight in the field in the night after a storm had to be done before life could begin again.

Although we were acutely aware of how the hail could wipe out days of work and hard-earned dollars in fifteen minutes, we were much less concerned about the long-term toll of the rain on bare soil. Those hard rains—aptly nicknamed gully washers—were not very well absorbed by the sparsely vegetated soil. Cotton does not have a large root system that would hold a lot of soil, and the middles were weed-free, a source of pride but also an invitation to erosion because they were bare of vegetation. The more the fields eroded and the more organic matter washed away and was not adequately replenished, the less rain could be absorbed, causing more runoff, which accelerated the whole vicious cycle.

It wasn't until I was a teenager and given the job of cultivating cotton that I saw the sandbars in the draw and the brown water racing down the ditch for what they really were: evidence of soil erosion in our fields. I already hated the sandbars because I dreaded getting the tractor stuck in them—there was no worse embarrassment for me than getting the tractor stuck. With this evidence of water-caused soil erosion and the constant threat of wind-caused erosion, it is no wonder that I chose soil erosion as a subject for my FFA speeches and began studying the subject. It was during this research that I learned that the soil we lost from our farm had an impact beyond our fields—dirtying first Otter Creek, then the Red River, and then the Mississippi River, affecting people who had nothing to do with agriculture.

Surface Water Pollution

Disturbing the soil through tillage and leaving it to the elements is a persistent problem in agriculture that results in vast amounts of sediment polluting surface waters each year. Sediment can fill in canals, reservoirs, and harbors, raise streambeds, decrease water storage area, kill fish and other aquatic life, and even destroy coral reefs. Dirt in the water can increase the frequency and seriousness of floods, increase the cost of treating drinking water supplies, and diminish freshwater and marine recreation. Sediment is not directly harmful to human health, but it is very expensive. Annual "off-site damage" from soil erosion adds up to an estimated nine billion dollars (in 1986 dollars).[2]

When too much silt enters a stream, it affects the penetration of light, the temperature of the water, the conditions on the bottom, and the retention of organic matter. The results include high mortality rates for freshwater mussels living in gravel-bedded or sand-bedded channels and the death of salmon fry.[3]

Besides sediment, agriculture contributes pesticides, the residues of chemical fertilizers, and nutrients from animal manures to surface water (streams, rivers, ponds, and lakes). According to the Environmental Protection Agency, agriculture is the leading source of water quality impairment in lakes, ponds, reservoirs, streams, and rivers, and the third leading source of impairment in estuaries.[4]

Impaired water cannot support or only partially supports beneficial water uses—aquatic life, fish and shellfish consumption and harvesting, drinking water, swimming and other water-based recreation, and agriculture itself.[5]

Impaired water is, in other words, unhealthy water, resulting from an unhealthy system of agriculture. It is apparent that industrial agriculture, with its overapplication of agricultural chemicals, its careless handling of organic wastes, and its failure to adequately address soil erosion, is guilty of endangering an essential natural resource: water.

Impaired water is, in other words, unhealthy water, resulting from an unhealthy system of agriculture.

Statistics from one state, Florida, help tell the tale. In the late 1980s, Florida had over 616,000 surface acres of lakes affected by "major" agricultural pollution.[6] Agriculture runoff in the United States reportedly pollutes 100,000 river miles. Such bodies of water can take years or decades to purify themselves.[7]

Two of the most damaging substances that find their way from farmland into surface water are the nutrients nitrogen and phosphorus in the form of nitrates and phosphates. (Nitrogen tends to flow with water, and phosphorus, chemically bound to the soil, is carried with sediments as the result of erosion.) Both chemical fertilizers and animal manures can be the source of these nutrients. Putting too much fertilizer on a field or applying it at the wrong time means it is more likely to be lost in runoff and end up polluting water than to be utilized by the plant.

When present in excess, these nutrients can cause algae blooms, which eventually deplete the water of oxygen, causing fish kills and damage to other living things in the water. A case in point: the 7,000-square-mile "dead zone" in the Gulf of Mexico that cannot support most aquatic life. Many other less dramatic "dead zones," however, occur in ponds, lakes, and streams in farm country every year.

Is agriculture really to blame for this? Agricultural activities, primarily row crops and livestock production, account for over 80 percent of all nitrogen added to the environment. Fertilizer is the single largest source of nitrogen; in 1995, American farmers used twenty-three billion pounds of nitrogen fertilizer, primarily for production of corn and wheat. This represents a twenty-five-fold increase in total annual nitrogen fertilizer use in the fifty-year period between 1945 and 1994. Not surprisingly, nitrogen from fertilizer is considered by many to be the most important preventable source of nitrate contamination of water supplies. Animal manure is the second largest source of nitrogen added to the environment, accounting for thirteen billion pounds per year.[8] This is a long way from the ideal described by Sir Albert Howard. In Nature's farming, he said, "there is no waste; the processes of growth and the processes of decay balance one another; ample provision is made to maintain large reserves of fertility; the greatest care is taken to store the rainfall."[9]

According to the NRCS, the "average annual nitrate-nitrogen concentration in the Mississippi River has doubled since 1950" with farm runoff the main source. This fist of nitrogen from the Mississippi basin eventually punches down to the Gulf of Mexico and knocks out vulnerable aquatic life, not to mention wounding the billion-dollar fisheries industry in Louisiana.[10]

This fist of nitrogen from the Mississippi basin eventually punches down to the Gulf of Mexico and knocks out vulnerable aquatic life . . .

Making the news in recent years is the fish-killing form of the marine microorganism, *Pfisteria piscicida,* in North Carolina and the

Chesapeake Bay. The rather sudden increase of this dangerous organism has been blamed on nutrient pollution from CAFOs, where tens of thousands of chickens and hogs are raised in cramped conditions in large buildings. Their waste, finding its way into the water, is a potent pollutant. As environmental activist Robert F. Kennedy Jr., put it, *Pfisteria* "inflicts pustulating lesions on fish whose flesh it dissolves with excreted toxins." This "cell from hell" reportedly killed a billion fish in just one incident.[11] In addition to killing fish in an ugly manner, *Pfisteria* toxins can also have adverse effects on the human nervous system, such as memory loss, respiratory problems, and skin rashes.[12]

Then there are the more mundane incidents: the increasing number of fish kills in recent years from improper handling of hog wastes from CAFOs. In Iowa alone, eighty-nine violations of waste handling rules have occured—either spills or improper land application of liquid waste since 1992—many resulting in fish kills.[13] The problem occurs wherever these CAFOs have been established.

Read the Label

Farm wastes often include leftover agricultural chemicals and chemical containers. Many chemicals have stringent storage guidelines—such as storing at certain temperatures, only in original containers, etc. One way to avoid dealing with this problem is to use fewer chemicals and thus lower the potential for spills, as well as disposal headaches. Another good strategy is to buy the right amount for the job; then there is no need to deal with leftovers. Another strategy is to choose the least toxic chemical available.

Farmers who use chemicals must dispose of leftovers and containers properly. The best advice, and one that farmers have been told time and again, is to follow the advice on the label for disposal. Improper disposal can result in groundwater contamination, fish kills in streams, and human health problems. The wrong (though unfortunately, common) thing to do is to toss leftover chemicals into farm "dumps" to slowly seep into streams and wells.

Some other farm wastes, such as old tires and batteries, have also been routinely dumped in the farm trash heap. Now these can be recycled, as can used motor oil. One can construct a heater for the barn or workshop that burns used oil.

How well a farmer manages farm waste is being scrutinized more than ever before. Some banks are now requiring an environmental audit on land on which they are lending money. An environmental analyst surveys the farm, checking for such hazards as leaking underground fuel tanks, chemical spills, and wastes, and then makes an environmental liability report to the bank.

Farm pesticides can also pollute streams and lakes. It is easy to imagine how herbicides and insecticides might damage aquatic environments and food chains. Insecticides can kill aquatic insects that other aquatic organisms rely on for food, and herbicides may damage phytoplankton, the base of the aquatic food chain.

Montreal veterinarian Martin Ouellet has concluded that the culprit causing deformities in many frogs is pesticides. He compared the rate of deformity in frogs on agricultural land untouched by pesticides for decades to those on working farms that use pesticides and chemical fertilizers. On the former, he found one frog in 100 deformed; on the latter, an average of twelve in 100 were deformed, and also suffered more severe deformities.[14]

Groundwater Pollution

Another danger from overuse of pesticides and fertilizers is the contamination of groundwater. Half of the U.S. population—almost all of those in rural areas—draws water from underground aquifers to drink. In western Oklahoma when I was growing up, the spring rains were essential not just for the crops but because everybody back then had a well. Everybody talked about their neighbor's water: Who had the best water, the deepest well, the shallowest well, and whose well had gone bad were prime topics of conversation. Besides barn raisings, I remember best the well and cistern diggings. I helped dig my grandfather's cistern—it was hard red clay all the way down, about fifteen feet. Like most such systems, it was linked to the guttering on the house so that the soft spring rain could be captured and stored there for summer use. Some people had both wells and cisterns; in those cases, the cistern was used as a backup in case the well went dry during a drought.

As we sprayed and fertilized, the possibility that these chemicals might leach into our water well was not something we worried about. This was long before people thought much about the purity of their drinking water, long before the advent of readily-available water filters that fit on the tap. Although toxic waste dumps, cesspools, landfills and septic tanks contribute their share of wastes to groundwater, agricultural chemicals contribute the most in sheer volume and affect the greatest area.[15]

We never tested our water; and, even today, one American in six draws drinking water from a private well or spring, many not tested for water quality.[16] Excess nitrates from fertilizer (and manure) can leach into groundwater and, in high enough concentrations, make such water dangerous to drink. (The EPA has set a maximum contaminant level [MCL] of ten parts per million for nitrate-nitrogen in public water supplies.)

At high levels, nitrates can be changed to nitrites by the bacteria in our bodies and reduce the oxygen-carrying ability of our blood, especially in fetuses, babies, and young children (also those being treated for peptic ulcers, persons with chronic gastritis, dialysis patients, and older persons). This is known as methemoglobinemia, or blue-baby syndrome. Infants suffering from the syndrome may show intermittent signs of blueness around the mouth, hands, and feet. They may have episodes of breathing trouble and some diarrhea and vomiting. In some cases, an infant has a peculiar lavender color but shows little distress. If blood samples are taken, they will be chocolate brown and won't turn pink when exposed to air. In severe cases, there is marked lethargy, excessive salivation, and loss of consciousness.[17] In two South Dakota cases, babies afflicted with this frightening condition drank water with high concentrations of nitrites (54 ppm and 150 ppm), but a case in Colorado involved an infant ingesting municipal system water containing only 13.3 ppm.[18] Nitrites can also form nitrosamines in the body, which are suspected of causing stomach cancer.

Just how big a threat these chemicals in water pose to public health today is a matter of debate. Some proponents of industrial agriculture claim that the threat is overblown. But it seems that those who drink from private wells, which includes many in rural areas, are at risk. In a study of well water for homes, the U.S. Geological Survey found in 1995 that 9 percent had levels of nitrates above safe drinking-water standards, significantly higher than a 1990 EPA survey of home wells, which found unsafe nitrate levels in 2.4 percent of wells.[19]

Often pollution from nitrates goes hand in hand with pollution from other farm chemicals. Pesticides can also pollute groundwater in agricultural areas. Groundwater is water that has percolated downward from the surface, filling the voids or open spaces in rocks. According to the NRCS, pesticide residue in groundwater seldom exceeds water quality standards.[20] However, according to the EPA, there is no known way to remove pesticide residue from groundwater or from people, for that matter.[21] So these contaminants are, for all practical purposes, permanently polluting deep aquifers.

Even today one American in six draws drinking water from private wells or springs, many not tested for water quality.

Others say that farm chemicals may be even more of a problem than government figures show. Environmentalists in twenty-nine cities in 1995 tested tap water for herbicides (weed killers) used in agriculture. The cities were located in the Corn Belt, in Louisiana, and in Maryland. The group found that cyanazine exceeded federal standards in more than a third of eleven samples. They found atrazine levels above health standards in 17 percent of all samples. (Both cyanazine and atrazine are thought to cause cancer in humans.)[22]

Because farm water wells are usually located near or on pastures and croplands, farm families are more likely to be exposed to dangerous pesticide levels or elevated nitrate levels in their drinking water than the general population. Farmers each season face many more real risks than most people in our society face, including the risk of financial ruin, not to mention injury and death from accidents with machinery and exposure to toxic substances. (Over the past twenty years, agriculture has become the most hazardous occupation in the United States).[23] Isn't it asking too much, then, to add unhealthy drinking water to the list of farming hazards?

WASHING THE WATER

Agricultural approaches to protecting water quality fall into two general categories: those approaches that focus on preventing the problem and those that attempt to solve the problem after it occurs. Sustainable agriculture focuses on approaches that avoid problems and will work over the long run. (For an illustration of management approaches that enhance and restore riparian areas, see Figure 4.1.)

FIGURE 4.1. Selected Management Opportunities to Maintain, Enhance, and Restore Riparian Areas

Cropping

Pasture/ Grazing

UPLAND ZONE	RIPARIAN ZONE	WATER COURSE	RIPARIAN ZONE	UPLAND ZONE
Conservation tillage	Enrichment plantings	Maintain instream flow	Fencing to manage livestock access	Grazing management
Crop rotation	Streamback stabilization	Manage dams and diversions to approximate natural flow variation	Enrichment plantings	Buffer strips
Pesticide management (IPM)	Buffer strips	Enhance aquatic habitat (long, weirs)	Streambank stabilization	Nutrient management
Buffer strips			Buffer strips	Set-backs and zoning
Nutrient management				
Set-backs and zoning				

Source: NRCS/RCA. "Riparian Areas: Implications for Management." Issue Brief 13, December 1997.

Precision farming is another way to prevent the problem of agricultural chemicals washing from fields. Working with a satellite, farmers apply fertilizers in the correct strength only to the places in a field where sensors attached to the tractor detect they are needed, thereby avoiding overapplication. Unfortunately, this kind of technology is expensive and out of the reach of small and medium-sized farms.

Pesticides, too, can be applied much more conservatively. Sometimes they can be applied at less than the recommended rate and still be effective. Banding pesticides—applying them carefully only to the crop row—is another way to make sure that excess pesticide doesn't get into the water. Adopting these kinds of practices can make a real difference. A computer simulation found that by using a combination of pesticide banding, conservation tillage, and reduced rates of pesticides, risk to fish could be reduced 77 percent.[24]

These are conservative, conventional approaches, and they are laudable. They work well in the short run. But they don't address the larger question of what will work in the long term as energy supplies get tighter. And they don't address finding ways to farm that would make pesticides and fertilizers much less necessary.

For example, growing cover crops is a way to keep soil in the field, thereby preventing the problem of sediment polluting the water. Cover crops can also reduce the need for pesticides by providing cover to beneficial insects and, if legumes are used, reducing the need to fertilize.

At the Kerr Center, we have tried a number of strategies to prevent water pollution. Since we raise mostly beef and just a few small plots of vegetables and fruits at the horticulture farm, we don't face the problems that farmers face growing crops such as wheat and corn season to season. But, even on a cattle ranch, there are areas that are prime for erosion, notably around water sources that are trampled by cattle. A typical cow, depending on the breed, can weigh 1,500 pounds—multiply that by 200 head and the word *trampled* hints at the effect. Vegetation is pummeled and the soil is exposed to the elements. We attempt nevertheless to keep these areas especially favored by cattle well-covered with grasses and clovers through seeding and by managing the grazing of the cattle so that no one area bears the brunt of their hooves for too long. Grazing pastures in rotation allows pastures to "rest" for periods of time and plants to grow densely.

The effect of cattle on aquatic life can also be a problem. When cows walk into ponds to cool off or drink, they trample tender vegetation that provides cover for fish and hatching grounds for the insects they like to eat. Cattle herds can also stir up the muddy bottom of ponds and burden the pond with waste.

At the Kerr Center, we have fenced off ponds and installed freeze-proof water tanks located below pond dams, gravity fed by pond water. This keeps cattle away from fragile pond banks. We have also funded projects around the state that improve water quality. One in western Oklahoma uses windmills to pump water to tanks scattered around the pastures. Another funds the installation of "limited access watering points"—generally, the pond is fenced off except for a single graveled walkway that provides access to the pond. The gravel gives the cattle firm footing and keeps the muddy bottom unstirred. A glance at the Environmental Stewardship Award Winners in each issue of *The Beef Brief* magazine confirms that around the country progressive cattle ranchers are adopting such strategies. There are clear payoffs to their efforts—cattle drink cleaner water, soil erosion is limited, and ponds last longer. A healthy pond is a wonder—home to a large number of plants and animals, from water lilies to bass, dragonflies to mallards. It can be a source of pleasure and pride on the farm.

Perhaps our most extensive water quality efforts have been our agroforestry projects. Agroforestry is, just as its name implies, a mixture of agricultural systems and forestry systems. One system of agroforestry, silvipasture, is the grazing of livestock and the growing of trees on the same land. It is a natural in the hills and mountains of southeastern Oklahoma, with its plentiful rainfall that can support many varieties of trees. In fact, most of the ranch was originally forested.

Agroforestry can be used to help reach several of the goals of sustainable agriculture, including improving water quality by reducing sediment and nutrients entering streams. Vegetation, such as trees, acts as barriers to runoff, either by slowing and absorbing water or catching sediment. These strips of trees at a prescribed width on each side of a waterway are called forest buffer zones. They are being promoted nationwide as a relatively cheap way to improve water quality. In fact, the USDA's National Conservation Buffer Initiative has a goal of establishing two million miles of buffers nationwide by 2002. Their hope is that farmers and ranchers will see the value of these buffers and adopt them voluntarily, heading off mandatory controls that could be imposed by government.

Contrary to the West African proverb which claims that "filthy water cannot be washed," trees and other vegetation in buffer zones along streams and lakes can remove sediment and chemicals before they reach the water. Even buffers that have no trees—grass filter strips they are called—can trap 70 to 80 percent of the sediments and contaminants from field runoff.[25] Much of the research to date has focused on the "removal efficiencies" of forested riparian areas and grass filter strips. Forest buffer systems are particularly good at removing phosphorus that has bound to sediment. They are also good at removing nitrogen, especially in areas with shallow groundwater.[26]

. . . trees in buffer zones along streams and lakes can remove sediment and chemicals before they reach the water.

Basically, these strips of vegetation slow runoff and retain sediment and other pollutants that would otherwise flow right into the water. Both woody and nonwoody plants can use what would be pollutants—such as nitrogen and phosphorus—for growth. Microbes also may break down pollutants so that they are immobilized or rendered nonpolluting. How does it happen? In the soil, among tree roots and grass, soil-dwelling bacteria consume nitrates and other nutrients from excess fertilizer and manure that has leached into shallow groundwater.

The effects can be dramatic—researchers have recorded a drop in nitrate levels from 15 ppm at the edge of a field to less than 1 ppm after groundwater passed through a riparian forest buffer before reaching a stream.[27] Another study found that either a riparian forest buffer or grass filter strip reduced herbicide concentrations in runoff water from 34 parts per billion (ppb) at the field's edge to 1 ppb or less near the stream. A bonus for farmers—they can harvest the grass for animal feed or harvest portions of the buffer without compromising the buffer's integrity.[28]

The Kerr Ranch borders the Poteau River. Beaver Lake is a large oxbow lake of the river situated at the bottom of a rather steep hill that was originally covered with forest, but which had been cleared for pasture. Overflow from a creek had carved a gully fifteen feet deep, twenty feet wide, and one-quarter of a mile long, the biggest in the ranch's 4,000-plus acres. The lost dirt had been deposited in a delta in Beaver Lake. Our goal was to repair this gash, keep others from forming, and protect the lake from more sediment coming off the hill in sheets.

We did it by widening and flattening the gully, changing its shape from a narrow "V" to a wide "U." Then we planted it in grass and trees. Agroforestry specialist Tim Snell planted eighteen rows of trees on the contour around the hill and across the gully, slowing water as it flows downhill. The days of gullies are over on this hillside. Instead of going into the lake, the sediment is caught by the trees, and we are now finding ridges of soil forming between the rows of trees.

The effect is more pronounced in another area of the ranch. On a wet, thirty-nine-acre flat piece of bottomland next to the Poteau River, we have planted almost 10,000 red oaks, ashes, and bald cypresses. Each time the area floods, the trees catch the sediment that flows in with the water. Now there are one to two inches of soil built up in the lower places that would otherwise eventually have added to the pollution in the Poteau River, and then the Arkansas, and eventually the Mississippi and the Gulf of Mexico. Along the way the pollution would make it more costly for cities that draw on these rivers to deliver clean water to their inhabitants.

Besides cutting pollution, trees absorb and use water on the farm that would otherwise be lost downstream—they are in effect, gigantic living water jars. Snell selected tree plantings that can also be eventually harvested for commercial uses or ranch use as fence posts.

In the last couple of years, we have begun fencing off many of the creeks and draws that run across the ranch. Once fenced off from the cattle, the waterways immediately begin to change—in fact, to heal. There are places on the ranch where one can stand and look one way and see an unfenced stream and the other way and see a newly fenced-off portion. The difference is striking—first in the quality and lushness of the vegetation in the fenced-off areas. Tall, thick grasses, both native and introduced species, are thriving along with forbs (broad-leaved plants such as sunflowers and ragweed). In addition, the stream course and streambed are changing, becoming after only one year, deeper and more meandering, less silty. New aquatic vegetation, not seen for years in these streams, is establishing itself—a key to establishing a viable food chain. The manager of the project, David Redhage, has been amazed at how quickly the area is reestablishing its natural life cycles. Although we have planted a selection of native trees adapted to our climate and soil in the fenced-off areas, other trees also have been popping up, seeds carried in by birds or on the wind.

One barrier to establishing buffer strips is cost. However, the easily moved and effective electric fence has substantially cut the investment one used to have to make, when the best option was a permanent barbed-wire fence. And there are currently a few government programs that provide money for buffer strips. Riparian buffer zones can also be used in a rotational grazing plan so that they are not completely removed from production. And because conservation buffers can effectively counteract erosion, their long-term impact on productivity is nothing but good.

DON'T DRINK UP THE POND

Most Americans take water for granted—it flows so easily from the tap into our glasses. But fresh water is, in reality, relatively rare—97 percent of the water on earth is in oceans and estuaries and not available for drinking or irrigation, and much of the remaining 3 percent is trapped in glacial ice. Fresh water, like soil, is essential to all animal and plant life on earth. Our bodies are largely made of water and we need to ingest two and one-half quarts of water per day, supplied from food and drink. Water is the main component of many foods; even meat, which seems so solid, is more than half water.

Plants draw water from the ground through their roots up to leaves where it transpires; transpiration being the plant equivalent of perspiration. One acre of corn in summer can reportedly transpire 3,000 to 4,000 gallons of water daily. Water is in the pore spaces in soil and, being an excellent solvent, holds nutrients and dissolved minerals and other substances absorbed by plant roots and essential to plant growth. Water in the soil also helps make the environment hospitable to microorganisms, which are so important for healthy soil.

Crops can be grown under dryland conditions (depending only on natural precipitation) or under irrigated conditions. In either case, adequate soil moisture is necessary for optimal growth. Of course, water can be detrimental to crops if the soil becomes overly saturated and literally suffocates roots by decreasing the oxygen in the soil. It is therefore in the best interest of the farmer to keep the right amount of water available to his plants. The best way to do this is to improve soil structure. By adding organic matter to the soil, it becomes more spongy and able to absorb water. Conservation tillage and green manures are two ways to add organic matter to topsoil and to improve soil structure. Other ways to give plants the water they need: cover crops slow down water movement, so that the water soaks in and is retained in the subsoil; surface litter, such as found with conservation tillage, greatly increases water infiltration at the soil's surface.

The United States uses vast amounts of water. Per capita daily water use in the United States is 1,400 gallons (which includes irrigation, mining, and manufacturing, as well as domestic use). The worldwide average is 475 gallons, but the range is huge, from five gallons per day in Haiti to 3,000 gallons per day in Iran.

Even in the rain-blessed Poteau River Valley (forty-five inches per year), water can no longer be taken for granted. The population is growing here, as it is all over the United States, and with it the demand for clean water. Half of all Americans use groundwater for domestic needs. At the same time, farmers depend on it for irrigation. Two-thirds of groundwater is used for irrigation. In the West, 90 percent is used for irrigation. Competition for water will inevitably develop.

To be sustainable, water use should not exceed the annual precipitation it takes to replenish surface water and groundwater sources, but it does in some areas of the United States. Until World War II, streams and rivers were the source of irrigation water in the Mountain and Pacific regions of the United States. Since then, wells drawing on groundwater have been the main source of irrigation water. One hundred billion gallons of fresh water are used daily to irrigate crops in this country. Unfortunately, the rate of use exceeds the recharge rate in many areas, and groundwater levels have been dropping.

. . . the Ogallala water is being used faster than it is being replenished, and the result will one day be "serious economic pressure on the area."

The Ogallala aquifer is a case in point. The nation's largest, covering 10,000 square miles from Texas to the Dakotas, it is a major source of water in the Oklahoma panhandle, where more than 2,000 irrigation wells have been drilled. This water supports farms that grow feed for the large cattle feedlots and, more recently, hog-feeding operations, in the arid (twenty inches of precipitation per year) area. These crops could not be grown successfully in the area without irrigation.

Unfortunately, the Ogallala water is being used faster than it is being replenished, and the result will one day be "serious economic pressure on the area."[29] The aquifer could be depleted within decades. In the meantime, while water levels drop, pumping the water up becomes more expensive.

Obviously, conservation is needed; irrigation is wasteful. In one traditional irrigation method, furrow irrigation, water is carried to the crops in a ditch; 40 to 60 percent of the water is lost before reaching the crop.[30] I know this from direct experience. Although the farm I grew up on in southwestern Oklahoma was by and large a dryland farm, there was one piece of cotton ground we rented from time to time that was irrigated. It was tempting to irrigate. As I described earlier, spring storms could be "gully washers" as the precipitation in southwestern Oklahoma is unevenly distributed throughout the year. In the month of August, when the highs often reach 100 degrees or more, not even two inches of rain fall in the average year.

Using an electric pump, we pumped water from two wells into a pond, then from the pond into irrigation ditches, using a tractor to run the pump. It was wasteful in almost every way. It took three days for the ditch to be saturated enough that the water would flow. Then the water flowing in the ditch, flowing to the field under those hot, dry conditions, evaporated quickly. The tractor ran constantly, using large amounts of fuel—although it was very cheap fuel. We used propane that cost eight or nine cents a gallon. Today, such a system would be cost prohibitive.

Unfortunately, irrigation was not the magic bullet. We found that the irrigated cotton almost inevitably developed more insect and disease problems than the dryland crop, and we would have to spray more pesticides. Every year it was a real question, even in those days of cheap fuel, whether it was worth it financially to rent this land. It would pay off every few years—when conditions were right for a bumper crop and prices were high. But other years it did not pay off. Government subsidies helped make the practice more attractive, though—the bigger your farm, the bigger the checks.

Efficiency and Beyond

Furrow and sprinkler systems of irrigation are the most common. A large system common in the Oklahoma panhandle is center pivot irrigation. Although more efficient than the ditch system, it still wastes 20 to 25 percent of the water pumped.[31] But a number of more efficient irrigation systems have been developed. Modifications to furrow systems capture runoff at the end of the field or control water flow. Low energy precision application (LEPA) is an efficient sprinkler system that can raise efficiency to 96 to 98 percent.[32]

Of the nation's cropland, 16 percent is irrigated. Almost half of this acreage is in Texas, California, and Nebraska. In Nebraska's Central Platte Valley they grow lush crops of corn, fed by nitrogen fertilizer and water pumped from the aquifer eighteen feet below the fertile silt loam soil. The bounty has a cost: almost 500,000 acres lie over groundwater that exceeds the safe drinking water standard of 10 ppm of nitrate-nitrogen. A major culprit of this unsafe water is nitrates leaching from the crop root zone into groundwater, a process accelerated by inefficient irrigation practices.

As a result, researchers from the USDA and the University of Nebraska have developed efficient "fertigation," where fertilizer is applied through irrigation water. One such technique is surge irrigation, which uses computer-controlled valves to apply water more uniformly along the furrow. The result: Half as much water as conventional furrow irrigation used, while maintaining corn yields. Another system uses sprinkler irrigation where nitrogen is "spoon-fed to plants" using readings from chlorophyll meters, which read the plants' nitrogen needs. After several years of sprinkler irrigation with improved nitrogen management, the water leaving the root zone was approaching 10 ppm, down from 32 ppm at the project's beginning.

This kind of research is valuable. But did farmers in the area adopt the new systems? Educational efforts by the University of Nebraska Cooperative Extension, the NRCS, and conservation districts helped reduce the water applied to the area by 10 percent, and the nitrogen by 20 percent.[33] Not spectacular results, but good ones. The question is: Are they enough, in the long run, to make a difference?

Drip irrigation, where water is applied very slowly to the root zone of the plant, is a system that many gardeners have adopted and that we have used on the horticulture farm during the often very hot and

Checklist for Farmers:
How to Conserve Water and Protect Its Quality

1. Conservation tillage methods
2. Increase soil organic matter
3. Cut chemical use
4. Use the least toxic chemical for the job
5. Establish conservation buffer strips
6. Protect riparian areas
7. Terrace
8. Farm on the contour
9. Strip cropping
10. Build ponds to catch sediment
11. Rotational grazing
12. Efficient irrigation
13. Grow crops adapted to the climate, soil type, and topography
14. Fence cattle out of waterways and ponds

dry late summer. It too is quite efficient, at 96 to 98 percent. Combined with a mulch to decrease evaporation from the soil's surface, drip irrigation helped us minimize water usage.

At our subtropical research station in Vero Beach, Florida, we use a drip system with a microemitter at each citrus tree. This is a departure from the common practice of flooding orange groves; we water just the tree root zone area and thus deliver the right amount of water to the tree at the right time and reduce runoff. Besides saving on pumping costs, nutrients and chemicals are not leached out of the grove into Florida's groundwater, which is already at risk from the many chemicals used in agriculture there.

Although efficient irrigation schemes are good, it seems to me that irrigation is most appropriate when it is used sparingly, as supplemental moisture, or when needed to avoid crop failure, rather than as a matter of course. And, most important, the water used should not exceed what is renewable. Raising crops such as corn in the Oklahoma panhandle is not sustainable—it takes too much water, which is not available there. So water is mined like gold, and one day it will be rare, like gold, there. Eventually farmers in the area will have to accept that their area is dry and drought prone and use the land to grow what is adapted to those conditions, which might be grass, as the land was originally a short-grass prairie. More research needs to be done to develop crops that are adapted to local conditions (see Chapter 5) rather than research on abating the impact of growing the wrong crops for local conditions, or abating the damage caused by the constant drive for more production. In the United States and around the world, new cropping systems need to be developed that significantly reduce or eliminate inputs such as fertilizer, pesticides, and irrigation water.

The next generation is depending on us to leave them a supply of clean, abundant water. In Oklahoma, as elsewhere, the rising cost of constructing water projects, the depletion of aquifers, the rising population, the demand for water, the shortage of reservoir sites, and demands for clean quality water are compelling reasons to conserve.[34] Perhaps we would all do well to take to heart a simple Native American proverb: The frog does not drink up the pond in which he lives.

MANAGE ORGANIC WASTES TO AVOID POLLUTION

It's April. The days are warming and so are the chicken houses of Le Flore County, Oklahoma. Le Flore County, named for a nineteenth-century chief of the Choctaw nation, is home to around 43,000 people and the Kerr Center. It is a lovely place, with green hollows and oak- and pine-covered hills. It is also the epicenter of the poultry industry in Oklahoma.

With temperatures no longer falling below freezing at night, there is no need for the heat-generating foot of chicken litter that has been accumulating on the floor all winter. It's time for spring cleaning of the two broiler houses owned by a typical chicken farmer in the Poteau River valley. In each of the long, low metal buildings, 20,000 chickens at a time are raised for six to eight weeks. Add to that some downtime for cleaning and maintenance, and in the course of a year the chicken farmer runs approximately five to six batches of chickens through each house.

The life of a chicken in a confined chicken operation is just that, confined. No sunshine or fresh air for her; no running after bugs or pecking the ground for seeds. Each bird is fed a prescribed amount of feed daily. She eats, grows, and gets heavier, while her waste constantly drops to the floor.

This piles up. The chicken industry rule of thumb is that each group of 1,000 birds generates a ton of litter per year.[35] At that rate, our typical chicken operation (with five batches) will generate about 100 tons per house—200 tons per year. Each cleaning—one in the spring one in the fall— therefore yields about 100 tons of litter.

The chicken industry rule of thumb is that each 1,000 birds generate a ton of litter per year.

The farmer owns this chicken litter. The farmer also owns (or more precisely is paying a large loan on) the chicken houses, which are enormously expensive and built to strict company specifications. But that is it—the chicks are supplied by the chicken company. The feed is supplied by the chicken company (paid for by the farmer). The companies (some of the most well known are Murphy Farms, Tyson, and around here, OK Foods) contract with farmers on a yearly basis to grow chickens for them. The farmers are not employees, neither are they independent, owning the chickens and selling them on a free market. They are something in between. However, one thing is for sure—the individual farmer is responsible for the litter generated in the operation.

What to do with such an enormous pile? Although the analysis varies, chicken litter can supply forty to eighty pounds of nitrogen, fifty pounds of phosphorus and forty-five to fifty-five pounds of potassium per ton,[36] a valuable amount of the big three nutrients. And since the chicken farmer also runs forty head of cattle on his eighty acres in the hills of southeast Oklahoma, he decides to load some of the litter into his manure spreader and spread it on his pasture—free fertilizer.

This is a good time to do it, he thinks—the weather forecaster has predicted rain at the end of the week. And before long, the spring rains will come every few days; the pasture, already wet, will get too muddy to drive on, as it does for varying times each spring. In farming, timing is everything. So he goes out on his tractor and drives over

How Chicken Litter Goes from Being a Fertilizer to Being a Pollutant

Waste generated by poultry raised in confinement operations is usually land applied. Under normal conditions, only a small fraction of the nutrients ever reaches the water. A large portion of the nitrogen is lost to the atmosphere. Plants take much of the rest, leaving only a small fraction of the original quantities of nitrogen to become a potential water pollutant. Phosphorus, although not volatile, often binds tightly with soil particles, leaving a small portion of the original quantities of phosphorus to become a potential water pollutant. However, because phosphorus levels in poultry waste are present in greater amounts than plants need in relation to nitrogen, phosphorus tends to accumulate on and near the soil surface. Phosphorus eventually becomes a water pollutant wherever poultry waste is used as a fertilizer year after year.

Source: Kevin Wagner and Scott Woodruff, "Phase I Clean Lakes Project: Diagnostic and Feasibility Study of Lake Eucha," Oklahoma Conservation Commission, February 1997.

the closely cropped "slick off the ground" Bermuda grass and lays down a big chunk of his manure pile—eighty tons on forty acres of pasture he fertilizes every year. At this time of year, the heat-loving Bermuda grass is dormant—it will be May before it really starts to grow. In this particular pasture, there are few cool-season grasses growing. When he gets back to the house, there is still a big pile of litter (though not as big) behind the chicken house, and the farmer feels like he has done a good day's work.

After the litter goes down on the ground, the microorganisms in the soil begin to work, breaking it down so it can be used as food for soil life. The next day, as predicted, it rains—a gentle rain, which helps to mix the litter in with the top layers of soil. Some time later, there comes a real gully washer—two inches coming down fast and hard in the night. Because the Bermuda grass has been overgrazed by the farmer's cattle, the grass is thin; there is not much in the way of cover to cushion the impact of the hard drops as they smack the dirt.

Before long, a broad sheet of brown water as wide as the pasture moves down the hill toward a small creek. It carries a heavy load—dissolved nitrogen compounds from the chicken litter, and small bits of soil carrying phosphorus bound to them. The pile by the chicken house door adds its load—brown water streaming across the yard and into the grader ditch. The water eventually makes it to the creek, then to the Poteau River, then to Lake Wister.

The Case of Lake Wister

Lake Wister is less than ten miles from the Kerr Center. It was created in 1949 when the Poteau River was dammed. It's a long, shallow lake—with a shoreline of 115 miles and a mean depth of only 7.5 feet—at its deepest only forty-four feet. Built primarily for flood control, it has evolved into a multipurpose lake—used also for recreation, and as the sole water supply for over 40,000 people, including the Kerr Center staff, in a three-county area. It's a lovely lake, lying as it does on the north edge of the Ouachita Mountains. Since 75 percent of its watershed is the forest of pines that cover these ridges, one might expect the lake to be relatively unsullied. But as the Oklahoma Water Resources Board summed it up: "Lake Wister has degraded since its impoundment . . ." In fact, Lake Wister's water is poor—it's turbid, which means murky or dirty because of sediment, and it's overloaded with nutrients such as phosphorus.[37] Some areas are covered with green slimy algae. The water tastes and smells bad. The cost of cleaning it so that it is drinkable has gone up and will continue to go up as the problem gets worse.

Wister, say water specialists, is eutrophic. The word eutrophic comes from a Greek word meaning "well-nourished." Eutrophication occurs when nutrient levels and biological productivity in a lake go up; a normal phase in the life of a lake as it slowly fills in to become a marsh and eventually dryland. But the process can be quickened by human activities. This acceleration, called "cultural" eutrophication, occurs when too much of a good thing—organic nutrients—flow into the water.

In a healthy system in balance, the breakdown of wastes in water begins when aerobic (air-breathing) bacteria, living on the oxygen dissolved in the water, digest the waste. This releases nitrates, phosphates, and other nutrients, which are absorbed by algae and aquatic plants and which stimulate their growth. Then it's up the food chain we go: zooplankton eat the algae, fish eat the zooplankton, and fish are eaten by other fish, birds, and animals. All along the chain waste is produced and organisms die. When they do, bacteria go to work, breaking them down for the next generation to use as food.

When too many nutrients come into a system such as Lake Wister, algae grows wild—an algae bloom. When these abnormally large amounts of algae die, as they must eventually, the aerobic bacteria needed to break them down increases. In the process, large amounts of oxygen are used up; the oxygen level drops and aquatic plants and animals die.

Why Lake Wister is declining is complex. Logging, mining, oil and gas exploration, and county roads contribute sediment, as do eroding fields and pastures. Wastewater treatment plants, urban storm water, septic tanks, and poultry processing plants contribute nitrogen and phosphorus. Nineteen point sources discharge into the Wister watershed, and it is estimated that they contribute about 12 to 16 percent of the phosphorus in the watershed, if they stay in compliance with average daily discharge rates. (Noncompliance has been documented, so the Water Resources Board suspects this number may underestimate actual point source inputs.)[38]

This said, it has been estimated that about 68 percent of the phosphorus comes from nonpoint sources.[39] The problem with nonpoint sources of pollution is precisely that: They are nonpoint. Unlike point sources of pollution, where one can stand in front of a drainpipe and test the water and know for certain if pollutants are being discharged, nonpoint pollution sources are diffuse. But it seems that fingers can justly be pointed at agriculture—including our chicken farmer and others like him. Over the past ten years, the vast increases in production of chicken waste, which mostly stays in the watershed, is thought to be one of the "predominant" sources of this nonpoint pollution.[40]

More than half of the upper Poteau River/Lake Wister watershed is in Le Flore County. In the past ten years, the number of "poultry production units" in the county has boomed. In 1994, there were thirty-six million broilers generating 55,000 tons of waste—110 million pounds annually. That's 3.3 million pounds of nitrogen and 2.8 million pounds of phosphorus. And production is expected to increase by 50 percent over the next few years, putting "additional stress" on the Poteau River and Lake Wister.[41] "Proper management of increased litter production has become a critical issue," said the Oklahoma Water Resources Board in its consensus report on the health of Lake Wister.[42]

I have watched as poultry production units have mushroomed in Le Flore County, often along the river. It happened so fast that nobody really understood what was happening. When it comes to building these houses two or four at a time, you don't realize how quickly they add up. We didn't realize that the production units were so numerous until suddenly we started having all these pollution problems.

. . . the vast increases in production of chicken waste, which mostly stays in the watershed, is thought to be one of the "predominant" sources of this nonpoint pollution.

For over twenty-five years, I have worked for opportunities for small farmers—the ones with limited money and land, the ones that these days are most apt to sign a contract with a large chicken company to raise chickens for them. I want people in Le Flore County to have the opportunity to make money from a farming enterprise. But I have reservations about the whole contract system, now used in both the poultry and hog industries, both because I think it is unfair to farmers and because it has caused environmental problems everywhere it has gone.

With mixed feelings, I have watched the chicken industry grow in Le Flore County. My feelings are mixed because I have friends and colleagues who are in the chicken business, raising chickens on contract. There are pluses to the contract raising of chickens; probably the best thing about it is that it's a way to remain on the land. Farming opportunities are limited everywhere—it takes large amounts of capital to get into and stay in farming. For many young people, it is almost impossible to get into, which is one reason the median age of farmers keeps increasing. In a county such as Le Flore, where most people have small acreages of land that are not particularly fertile, it is almost impossible to make it financially growing staple crops, such as wheat or soybeans, or, to make a living from raising cattle. Prices are too low and the economies of scale are against them. Horticultural crops are an option, but the markets are not well developed, and small operations such as these have not benefited from government programs or research or extension dollars. So it's no wonder that many decide to seize the opportunities offered by chicken corporations.

Farming opportunities are limited everywhere—it takes large amounts of capital to get into and stay in farming.

Then comes the pollution. In addition, a number of negatives are present in the contract raising of chickens, including many troubling questions about the fairness of contracts to farmers, debt burden, environmental liability, and concerns about how humanely these animals are treated. How quickly some rural citizens and communities have embraced contract production is indicative of the "desperation economics" of these communities.

Get Big and Pollute

A discussion of water quality problems leads inevitably to a discussion of other problems of industrial agriculture, of which confined feeding operations are a prime example. It also makes clear that how farmers raise the food we eat affects others far beyond the farm. An example is another lake in Oklahoma, Lake Eucha, where the water is as poor as Lake Wister's and for the same reasons. According to the Oklahoma Conservation Commission, the average annual nitrate concentration in the watershed doubled and the phosphorus concentration in the watershed tripled between 1975 and 1995. Within that time period, the chicken industry boomed in the watershed.

Eucha's woes have gotten more publicity than Wister's because it supplies a large percentage of Tulsa's drinking water. In the past couple of years, considerable friction has occurred between urban interests (Tulsa) and the interests of those who live in rural areas and have contracts with chicken companies. Tulsa has been spending increasing amounts of money cleaning up the water from Lake Eucha (an additional $100,000 per year) to make it drinkable and has pushed for changes in the way litter is handled in the watershed.

Concentration seems to be the main problem in this situation. There are too many chickens in the watershed. Chicken operations are concentrated in small geographical areas because companies don't like to deliver feed from their feed mills farther than in a sixty- or seventy-mile radius. The chicken litter generally stays in the watershed because farmers consider it free fertilizer. Although others outside of the watershed might like to utilize the litter, which can be a valuable organic fertilizer when used properly, litter is bulky and therefore expensive to haul, which makes it a less attractive option to those who might buy it.

Although farm chemical wastes are a problem nationwide, in Le Flore County and in areas of the country where confinement feeding of cattle, hogs, and chickens is common, animal wastes have become a major source of water pollution. In fact, animal manure is rated with commercial fertilizers and atmospheric deposition as one of the three primary nonpoint sources of nitrates in surface and groundwater, according to the National Water Quality Assessment done by the U.S. Geological Survey. And it is not just Oklahoma. A 1990 study of one river basin, the Apalachicola-Chattahoochee-Flint River, of Georgia, Alabama, and Florida, found that more than half the "nutrient load" of phosphorus and nitrogen in these rivers came from poultry manure.[43]

Raising large numbers of hogs in a confinement feeding operation can also generate a lot of manure that can become a pollutant. One hog, for example, can generate almost ten pounds of wet waste per day, of which nine pounds is water. If a farm has 2,000 hogs (small in today's market), this quickly adds up.

Hog CAFOs have become big polluters in recent years. The hogs spend their lives confined in small cages indoors, and their waste is stored in lagoons. At some point, the liquid from the lagoons is removed and often sprayed onto farm fields or pastures nearby as fertilizer.

The problem: lagoons can leak. Waste can overflow when it rains too much. (This happened to East Coast hog farms as a result of hurricanes in 1999.) Land can be overfertilized and waste can runoff. Equipment can fail or humans can make errors. These problems, unfortunately, have become all too common. The result is polluted water.

The problem is made astronomically worse by the size of some of these operations. A new hog CAFO planned for the Oklahoma panhandle will house about 250,000 hogs.

Not surprisingly, the scale and frequency of spills and other mishaps have become scandalous. In 1995, an eight-acre lagoon in North Carolina burst and spilled thirty-five million gallons of animal waste into the New River. The spill killed ten million fish and closed 364,000 acres of coastal wetlands to shellfish harvesting. In 1996, forty spills contaminated rivers and killed 700,000 fish in Iowa, Minnesota, and Missouri. In 1998, a 100,000 gallon spill into Minnesota's Beaver Creek contaminated the creek and killed close to 700,000 fish.[44]

. . . the scale and frequency of spills and other mishaps has become scandalous.

However, it is not just the big boys who are fouling the waters. Our hypothetical typical Le Flore County chicken farmer is a case in point. In attempting to manage the waste from his chicken operation, he unwittingly did a number of things wrong. He did not test his soil to check levels of nutrients. He therefore could not assess whether or not his pasture plants needed, in particular, more phosphorus. (See box "How Chicken Litter Goes from Being a Fertilizer to Being a Pollutant.") He spread it on a pasture without actively growing vegetation that could have taken up and used the nutrients. He allowed his cattle to overgraze the pasture so that it would erode more easily in a rain and carry phosphorus from the manure with it. What he didn't spread, he left in an uncovered pile, which leached into the bar ditch with the rain. Repeat this scenario enough times and the results can be seen in Lake Wister.

Whereas in the past many small farms raised smaller numbers of chickens and their waste was easily utilized on the farm in a way that didn't pollute, industrial chicken operations generate huge amounts of waste and endanger water, an essential natural resource. Therefore, managing farm wastes and protecting water quality are closely related goals. And although managing wastes has long been a problem for farmers, the area has not been benefited from the research and development dollars that production agriculture has, and it has been understressed in traditional extension programs. Another problem is the unwillingness of the chicken corporations to accept responsibility for their part in polluting the water. And perhaps most important, the corporate buck-passing—in this case pollution-passing—to the individual, financially strapped contract farmer, who gains the responsibility and liability for the waste that is generated by chickens that are actually owned not by the farmer, but by the corporation.

DON'T WASTE THE WASTE

The goals of sustainable waste management are to minimize the amount of waste and its toxicity, use it as a resource when possible, and, if not, dispose of it in an environmentally responsible manner.

Organic wastes are likely to be viewed as resources these days. A case in point is wheat straw, which used to be burned in the field, and is now often left there instead to deter erosion and to mulch growing crops. Although animal manures can pollute if applied improperly, they can also be very valuable fertilizers. Chicken litter is a valuable source of nutrients and organic matter, which contribute to soil structure, health, and tilth. Chicken litter also releases nutrients more slowly than chemical fertilizers, not shocking plants and soil as synthetics can, and providing nutrients over a longer period. In this way, it is a more natural fertilizer. Organic wastes also increase levels of beneficial microbes in the soil.

Although animal manures can pollute if applied improperly, they can also be very valuable fertilizers.

The use of organic fertilizer is in line with Sir Albert Howard's description of Nature's farming: "The mixed vegetable and animal wastes are converted into humus. And Mother earth never attempts to farm without livestock."[45]

It is very likely that when chicken farmers take actions that result in pollution, they do so at least partly out of ignorance. In Le Flore County and likely in other places, the poultry industry has expanded so quickly that educational efforts have not kept pace. As the spreading of chicken litter has been largely unregulated, it has been up to the individual farmer to get information about how to dispose of the waste correctly. Until recently, the poultry companies in this area did not supply much information to their contract farmers about waste disposal. Such information is available from the NRCS and extension, but usually there is no direct financial incentive for farmers to follow what are termed "best management practices" (BMPs) by the NRCS in regard to managing animal waste.

In 1994, the Oklahoma Conservation Commission, in cooperation with a number of county, state, and federal agencies, launched the Poteau River Comprehensive Watershed Management Program. The program included a number of initiatives: stations to monitor stream water quality, a program to monitor water quality in the lake, an educational program, and "water quality incentive" payments funded by the USDA Water Quality Initiative Program. Kerr Center staff member Alan Ware served on the administering committee. The overall objective of the program was to reduce loading of nitrogen, phosphorus, and sediment into Lake Wister and the streams that feed it.[46]

The project chose one watershed, Hog Creek, in which to concentrate its efforts and which would serve as a demonstration watershed. The committee sponsored a fish fry in the community to explain the program to residents, and it was well received. A major feature of the program was the cost sharing the government offered to farmers if they agreed to adopt certain "best management practices"—for example, building a cake-out shed. Cake-out is both a substance and a process; the farmer does a cake-out when he cleans out the top layer of litter around feed bins and waterers. This is the litter that has absorbed water and has caked. This is usually done between batches of birds—about every six weeks. Having a shed built to NRCS specifications to store this litter will keep it dry and will remove some of the farmer's incentive to spread it immediately on pastures or fields, which can be at the wrong times. The sheds also included a built-in chicken composter for dead birds, removing them as a source of pollution. Producers could get 60 percent of the cost of the facility paid for by the program, if they agreed to use it for ten years after installation.

Our hypothetical chicken farmer would have benefited from such a cake-out house, as well as from other incentives to help him manage his waste better. He might have received money to help pay for seeding cool-season grasses, such as fescue and rye, in his Bermuda grass pasture, so that if he spread litter on his pasture in April, there would be plants to benefit and absorb nutrients. He might have received money to fence off his creek, letting vegetation grow to make a buffer zone that will catch sediment and nutrients before they can enter the creek. Fencing the cattle out would also preserve the creek banks from being broken down by cattle hooves. He might even have received money to set up a rotational grazing system, so that his land would not be overgrazed and subject to erosion. In addition, he might have gotten a one-time per-acre payment for disposing of his waste properly—at the correct rate and time. A NRCS specialist would take a soil test and then recommend how much litter to apply to meet the needs of the particular area.

Various best management practices were shown off on annual field trips and so functioned as educational demonstrations. These field trips also served to educate the public about water pollution in general—on one such trip, participants were shown a healthy creek and a polluted creek's health. Water specialists explained what they look for in assessing a creek: the condition of the bank—is it stable or not, the number of trees, and the amount of shade; the amount of sediment filling up spaces in the rock; and what insects and fish are present. Besides field trips, workshops were held to teach producers how to write pollution prevention plans. Although the overall objective of the education/demonstration program was to teach producers how to properly use poultry litter as a fertilizer, funds were also available to help pay for replacing primitive septic systems, which were leaking into the watershed, and to fund pasture improvements.

Voluntary Strategies for Poultry Waste Management

Short-Term Strategies

- Research.
- Identify areas within the watershed that contribute a disproportionate share of nutrient load to surface waters.
- Identify best management strategies that will work in the watershed.
- Education—Conduct intensive farmer training and education activities about proper waste disposal. Conduct mass awareness campaigns in watersheds that are at risk.
- Encourage industry to support such activities.
- When possible use the USDA's Environmental Quality Incentives Program (EQIP) as a vehicle to accomplish goals.

Long-Term Strategies

- Research ways to increase the value of chicken litter and slow the release of its nutrients.
- Encourage sustainable approaches to raising poultry.
- Adopt as policy whole-farm, watershed, and sustainable approaches to land use.

In three years, the program contacted between 4,000 and 5,000 people through conferences and farmer meetings. Twenty-six contracts were written in which participants agreed to adopt best management practices and were given cost-share assistance.

Although this program by itself could not clean up Lake Wister, it is a first step. The cost-share funds are a carrot to get producers on the right track—a first step in better educating the community, in helping people make the connection between the dirty water in the creek their children play in and their own and their neighbors' actions.

In 1998, the Oklahoma legislature passed a bill that established, for the first time, regulations on confined animal feeding operations. They regulated both hog and chicken operations. Oklahoma had been known for its lack of regulations on CAFOs, but this bill changed that, putting Oklahoma "on the cutting edge" of CAFO regulation, as one observer put it. Kerr Center Rural Development and Public Policy Director Michelle Stephens served on the task force that investigated CAFOs in Oklahoma and recommended policy changes.

On the hog side, the law requires pollution prevention plans from hog operations, prescribes setbacks from residences depending on the size of the operation, requires leak detection or monitoring wells, and also sets a number of requirements for lagoons. The rules for chicken operations include requiring poultry growers to take nine hours of waste management training in the first year and three hours every year after that. They also require poultry growers to have a management plan that includes soil and waste testing periodically to determine the levels of nitrogen and phosphorus present, from which they will determine application rates. The legislature also set up a poultry waste transfer fund to encourage the transfer of waste out of watersheds that have too much phosphorus. Also part of the poultry bill is the prohibition of land application of poultry waste in certain watersheds if most recent soil tests indicate a phosphorus level greater than allowed by national standards.

Regulations can be an effective tool to reduce pollution, particularly when employed against large corporate hog operations. But against farmers, particular difficulties are involved in employing such an approach. I don't think most country people would favor heavy fines on neighbors for spreading chicken litter incorrectly, say on a pasture already saturated with phosphorus. Most rural residents strongly oppose interfering with a farmer's livelihood, especially considering the state of the farm economy. This was apparent twenty years ago, when there was a brucellosis problem in Le Flore County. Brucellosis is a serious cattle disease, and everyone knew whose cows were infected and that they were being sold illegally. However, the guilty parties were never prosecuted because of community sentiment against such interference.

I believe the situation would be the same if strict regulations were applied to farm waste management. Far better would be a change in the hearts and minds of Le Flore County farmers.

It is possible that programs such as the Poteau River watershed management program may serve as a forerunner to a more significant grassroots movement in the future. In my experience, much more is likely to happen when there is a change in consciousness—when people start talking about dirty water or soil erosion in the cafés over morning coffee and biscuits. When people join together and approach government agencies for help, rather than the other way around, change is likely to be more lasting.

What is likely to bring about such change? A crisis probably would. I imagine if Lake Wister got so foul smelling it kept people awake at night, or if the fishing declined significantly, people would take action. There is quiet talk about just abandoning Lake Wister and building a new lake by flooding a beautiful mountain valley. When land and/or money is needed for that, there may be some protests. As it is, water rates climb slowly as the water gets dirtier and it gets more expensive to make it drinkable, and we accept the gradual change.

. . . much more is likely to happen when there is a change in consciousness—when people start talking about dirty water or soil erosion in the cafés over morning coffee and biscuits.

I hate to think this is true, but we might be like the complacent frog in the legendary frog-in-the-boiling-water experiment. In this demonstration, the pot of water the frog sits in is heated slowly. As the temperature increases, the frog adjusts to it. By the time the water boils, the frog is cooked before it thinks to jump out.

It's my hope that we won't be like that poor frog; instead, we'll jump out of our familiar yet dangerous situation before we are, as it were, cooked. I hope that education will work and that farmers will internalize an ethic of stewardship.

One barrier to this happening is that farmers in many countries, including the United States, are unable to shoulder the full financial cost of environmental conservation. Offering them an array of green incentives could contribute to reducing pollution. These incentives might encourage farmers to try raising hogs and chickens in ways that are nonpolluting. A few pioneers around the country are trying these alternative systems, which are often more profitable than contracting with a large corporation. Ways of making animal waste into a valuable resource exist, without the negatives of CAFO systems, and they look to be more sustainable economically, too.

Smaller Is Better

Two viable alternatives to the contract growing of confined chickens are pastured poultry and free-range poultry. Both allow small producers to make a living on their farms, and they have considerable environmental advantages. Pastured poultry has an effective spokesman in Joel Salatin of Polyface Farms in Virginia, who developed it and has written and spoken extensively on the subject.

In the pastured poultry system, chickens are kept in portable pens that are moved often around a pasture. The chickens stay on clean ground, able to eat bugs and grass as they like. It is a humane system, and one that offers the farmer independence and a good living through direct marketing of the birds—$25,000 to $30,000 in six months on twenty acres. These farms are small, growing many fewer birds than a contract operation. The investment, too, is small—the price of a medium-sized tractor rather than the several hundred thousand dollars a contract chicken operation requires to establish.[47]

From a water quality standpoint, the system is good because production is seasonal; the birds are kept only on growing pasture, so the nitrogen they excrete is used immediately by the actively growing grasses. It is efficient nutrient cycling, says Salatin. He advises the use of cattle in a rotation to follow the birds and fully utilize the forage, so the nitrogen is converted quickly from grass to chicken and beef. Pastured poultry avoids overloading the environment, such as when large amounts of litter are applied to pastures a couple of times a year. The water supply is protected.

. . . the birds are kept only on growing pasture, so the nitrogen they excrete is used immediately by the actively growing grasses.

Checklist for Farmers:
How to Manage Organic Wastes to Avoid Pollution

1. Test soil and apply organic wastes to pastures and fields only when nutrients are needed.
2. Compost dead chickens.
3. Compost litter.
4. Shelter piles of litter from rain.
5. Apply litter to actively growing vegetation.
6. Grow chickens free range or on pasture.
7. Hoop house production for hogs.

Too, if the farm does its own slaughtering, the numbers of birds processed at one time is small. Compare this to a commercial processing plant which might process 100,000 birds per day, and which "then has tractor trailer loads of offal, feathers, blood and millions of gallons of processing water to treat." Salatin says they use half the water a commercial plant uses to process a bird, and the water is used in the garden or pasture as another organic fertilizer.[48] The chickens are sold directly to consumers.

Free-range poultry systems are a bit different than pastured poultry in that the chickens are given a larger area to roam, but are not moved as often. However, they have many of the same advantages for the producer, including independence and an opportunity to raise birds in a humane manner. But with them, too, waste is deposited continually where it can be utilized by growing forage. Ideal for the small-scale grower and direct marketing, this method was popular in the United States before the 1960s and continues to be used in Europe today.[49]

There are alternative hog production approaches too that are popular in Europe. They too are much better than confined operations from a water quality standpoint. Hogs can be efficiently raised in deep-bedded hog houses or in deep straw in hoop houses (hoop houses are hooped-pipe Quonset-shaped structures covered with fabric tarps). The advantage from a water quality standpoint is that the hog waste is absorbed by the deep layers of straw in these structures and begins to compost, instead of being stored in a noxious lagoon that might leak. It also eliminates the need to spray liquid hog manure onto fields and pastures, which is potentially polluting to water. (The odor associated with this practice can ruin the quality of life for anyone whose home is nearby, not to mention property values.)

In the deep straw systems, the bedding is converted into a solid fertilizer, which has little odor and can be spread on fields. Properly managed, the hog waste does not seep from the straw into the ground beneath it—providing enough straw to cover wet spots is essential. Minnesota farmers Mark and Nancy Moulton, who researched the risk of nutrient leaching into and through such bedding and into the subsoil, found the ground dry under the bedding in their hoop houses.[50]

Other sustainable nonconfined systems put pigs on pastures or fields to forage, taking advantage of their ability to utilize numerous forages and crop residues and reducing the need for off-farm feed and feed supplements. Manure is spread about the fields as the animals forage, so lowering the need for lagoons and liquid waste to dispose of.[51] The Kerr Foundation developed a low-investment, nonconfined swine system in 1985 that allowed hogs to range on pastures seeded with high quality forages such as clovers and small grains such as rye and ryegrass.[52]

The key to managing animal waste effectively is the development of small-scale, water-friendly systems scattered across the landscape, not industrial-style operations concentrated in one watershed. Bigger is not better—not for farmers and, as it seems, not for water quality.

Anatomy of a Manure Spill

This is a story with many vital statistics: 100,000 gallons of raw hog manure spilled; 690,000 dead fish; 18.7 miles of polluted stream. These numbers represent the results of one factory farm disaster on a typical creek in the Upper Midwest this summer. Unfortunately, they are not uncommon statistics in this age of industrial meat production. In rural America, initial shock is giving away to feelings of helplessness and apathy as the fish body count climbs, the rivers of manure rage, and the burgeoning banks of pollution line miles of streams. This apathy is the by-product of trying to digest and imagine facts and figures of astronomical proportions.

But this story takes a look beyond the sterile statistics of news reports and offers a down and dirty glimpse into what happened to one rural community's natural and human residents when factory farming went awry.

The Fish Aren't Biting

Mr. Barta has lived within a few hundred feet of the east branch of Beaver Creek all his forty-five years. So it was no big deal to walk the few hundred feet from his house to

(continued)

(continued)

the small stream with his eighteen-year-old son for an evening of fishing. But on Saturday, June 21, they couldn't even catch the lowly chub in the two- to three-foot water.

"They weren't biting," recalls Barta, who farms 640 acres of corn, wheat, and soybeans seven miles south of the western Minnesota community of Olivia. "I think they were too busy fighting for air."

That's an understatement. The chubs, bullheads, darters, minnows, and various other species of fish were literally gasping for their lives. By the time the Bartas arrived at the creek at about 4:30 p.m., the fish were at the surface, trying to get any oxygen they could. Some of the fish were clustered around places where fresh water was trickling into the creek, desperate for any new source of oxygen. Even more troubling was that crayfish, tough bottom-dwelling crustaceans that are seldom seen on dry land, were literally crawling up the banks to get away from the water.

This was a hard scene for the Bartas to take in. The creek runs a half-mile through their farm and Barta's son is the fourth generation of Bartas to go fishing in the creek, which had always been relatively clean even though it runs through one of the most intensely farmed areas in the country. But seeing fish struggle for oxygen was nothing compared to the scene of devastation that developed before the Bartas' eyes within a half-hour of arriving at the creek.

"Then all we saw is the dead ones," recalls Dennis. "The creek turned white with the bellies of dead fish. I knew something was wrong, but I didn't know what."

Hogging All the Oxygen

What was wrong was that tens of thousands of gallons of hog manure—Minnesota Pollution Control Agency (MPCA) officials estimate as much as 100,000 gallons—had washed into the creek roughly 10 miles upstream of the Barta farm. As nutrients such as nitrogen and phosphorus started to break down, they consumed all the dissolved oxygen needed to sustain life in the creek. In addition, the manure contained ammonia, which can be toxic to fish. The fish literally suffocated. The hog producer responsible for the spill has told the authorities that the manure escaped his operation sometime during the evening of Thursday, June 19. What the Bartas were witnessing was the climax of one of the state's worst manure-caused fish kills.

Source: Brian DeVore, *The Land Stewardship Letter,* Land Stewardship Project, Volume 15(4), 1997, pp. 1, 9.

Chapter 5

Think Local: Step 4—Select Plants and Animals Adapted to the Environment

Hurt not the earth, neither the sea, nor the trees.

Revelations 7:3

Titanic. It was a fitting name for such a big, powerful ship that could sail when and where it wanted to, even through the iceberg fields of the North Atlantic. The *Titanic* was promoted as a ship that could go where other ships didn't dare to go. Because of its superior engineering, the *Titanic* was thought to be unassailable, unstoppable, and unsinkable.

The *Titanic* was the product of a young, brash industrial culture that believed there were no limits to what was possible. Today, many people see modern industrial agriculture as the light-hearted embarkees saw the *Titanic* the day they set sail—big, powerful, unlikely, it is thought, ever to fail. It is an article of faith that industrial agriculture, like the *Titanic,* will be able to overcome any obstacle Mother Nature might throw in its path.

Of course, we all know what happened to the *Titanic*. When it sank in 1912, it took almost everyone with it, rich and poor alike. Its loss shook the confidence of a world embarking on what promised to be its most glorious century yet.

Agriculture does not have to repeat that sad story. I have no doubt that a deadly iceberg is out there with industrial agriculture's name on it. It could be any number of crises: the end of cheap fossil fuels and the products made from them, the wearing out of soils, widespread water pollution, or the concentration of the production and marketing of food into the hands of a few large corporations. Whatever the iceberg (or icebergs) might be, we can avoid the big crash by recognizing the dangers and correcting our course now, while there is still time to do it.

Practitioners and proponents of sustainable agriculture are doing just that. We all have the same destination in mind: a safe, plentiful food supply produced by independent farmers who receive a fair profit and who work to enhance, not degrade, the environment.

If industrial agriculture is the *Titanic,* then those of us working for a sustainable agriculture are trying to change her course before it is too late. Unfortunately, big ships move slowly. I often find myself doing some variation of "emergency measures"—talking to farmers' groups, testifying before governmental agencies—in essence, sounding the warning. At the Kerr Center and other sustainable agriculture organizations, and at the SARE program, we hope to change the course of a *Titanic*-sized institution called industrial agriculture.

On the other hand, it could be that industrial agriculture has already hit the big iceberg in the form of Mother Nature herself with her implacable laws. If this is true, we are, in reality, already sinking and need to move onto some new ships. They will be new kinds of vessels—smaller, quicker, lighter, and more maneuverable—ships that can avoid the icebergs and sail into new waters if need be. These new ships will not be titans powered by oil but sleek sloops powered by wind and sun. They will work with Mother Nature rather than against her, sailing with the currents and trade winds instead of against them. These new ships will be the flagships of a sustainable agriculture.

Now is the time for the blueprints to be drawn and these new ships to be built—for the ideas of sustainable agriculture to be given a fair hearing and the full attention of researchers and farmers. As fisheries biologist Kenneth Williams wrote in the Kerr Center newsletter a few years ago: "Sustainable agriculture is modern farming. It relies on all that has been learned about the ecological principles that govern life on our planet."[1]

This cannot be overemphasized, because some have tried to dismiss sustainable agriculture by claiming that it is an exercise in nostalgia, a move backward. This could not be further from the truth. Actually, it might be industrial agriculture that is a move backward, or at least a trip down a blind alley. The reality is that in the nearly

10,000-year time line of agriculture, industrial agriculture is just a blip—an experiment that is likely to fail, given what we have been discovering about agriculture and ecology.

AGROECOLOGY

"Sustainable agriculture is modern farming. It relies on all that has been learned about the ecological principles that govern life on our planet."

Just as industrial agriculture could not have arisen without the industrial revolution, so it is that many of the ideas of sustainable agriculture could not have arisen without the development of the biological sciences, particularly in the area of ecology. Sustainable agriculture is based on scientific ideas; ideally, it will live or die on the strength of its ideas. I say *ideally* because the forces aligned against it are trying to undermine it for other reasons, such as because it depends less on borrowed capital and because it is more conscious of resource conservation. This frugality threatens the money flow that has been established over the past forty years.

Today, we conceive of the earth as a biosphere, a sphere of life. Ecology is a science that investigates the relationships between plant and animal organisms and their environment. One ecological concept is that patterns of life reflect the patterns of the physical environment. For example, vegetation is influenced by climate and soil. Cacti grow in deserts, not in the boreal forest. Vegetation determines which animals live in an ecosystem. Pandas live in China rather than Oklahoma because they eat bamboo rather than acorns. Going up the food chain, the kinds of predators, too, are determined by the kind of prey available. The plants and animals that thrive in a given ecosystem can be said to be adapted to it.

Another ecological concept is that the plants and animals of a given area form a loosely organized group—a community. A grassland and a forest are two different types of communities. These, in turn, can contain smaller, more specific communities—such as the tallgrass prairie type of grassland and the oak-hickory type of forest.

Both competition and cooperation occur among the organisms in a given community.

In every ecosystem, whether it is a tropical rain forest or tallgrass prairie, the biological community exchanges matter and energy with its physical environment. It is self-sustaining. A natural ecosystem is powered primarily by the sun; it is also fed by nitrogen in the air and soil, and minerals in the soil that come from rocks that have weathered. Organisms of three types—producers, consumers, and decomposers—promote energy exchange is an ecosystem. Green plants make their own food through photosynthesis, using energy from the sunlight and water and carbon dioxide from the environment. Plants are producers. Animals that feed on plants or other animals are consumers. The bacteria and fungi that break down the bodies of both the producers and consumers after they die are the decomposers.

The farm can be looked at as a special kind of ecosystem, an agroecosystem. This way of looking at a farm was new to me when I first encountered it fifteen years ago, as indeed it is for most of us trained in agriculture in the 1960s and before. We of that generation are more likely to think of a farm as an agribusiness rather than an agroecosystem.

Depending on the stresses they are under, agroecosystems may or may not be sustainable. An agroecosystem is more apt to be sustainable if it can utilize nature's cycles. Of course, this is not an entirely new idea. In this century, those of the Nature School realized the importance of nature as a model for agriculture. Earlier peoples, such as the ancient Greeks or Native Americans, also seem to have been in tune with this idea. The Greek word *agrios,* meaning wild, is akin to their word *agros,* meaning field.

As an agroecosystem, a farm can be evaluated using ecological concepts. Fields of crops are influenced by climate and soil just as prairies are. And there are other parallels between natural ecosystems and agroecosystems. One can think of a field of crops—say a field of cotton in Kiowa County, Oklahoma—as a community of cotton, weeds, insects, earthworms, mice, soil microbes, hawks, and so on. It contains many fewer plants and animals than the mixed-grass prairie it replaced, but it is a community nonetheless.

Depending on the stresses they are under, agroecosystems may or may not be sustainable. An agroecosystem is more apt to be sustainable if it can utilize nature's cycles.

It is also true that in a field of cotton there are producers, consumers, and decomposers. The cotton plants are the producers. Anything that feeds on cotton bolls (such as the boll weevil) or harvests the bolls (such as the farmer) is a consumer. The tiny creatures who break down the postharvest cotton stubble to humus are the decomposers, recycling the nutrients in the stubble back to the earth to help feed the next plant that grows there.

However, important differences can exist between natural ecosystems and agroecosystems. Agroecosystems have been described as "semidomesticated," falling somewhere between ecosystems that have experienced minimal human impact and those under maximum human control, such as cities.[2] The human race first created agroecosystems when our needs exceeded what natural ecosystems could provide. Thus, people went from being hunters and gatherers, dependent on the natural ecosystem for food, to being farmers and herders, dependent on the agroecosystems of their own design.

Agroecosystems will always be disruptive to natural cycles because they are controlled by humans. However, the difference between a sustainable agroecosystem and one that is not is in the degree of disruption. With modern industrial technology, the disruption can be massive and quick. Give me a powerful tractor and a big plow and I can turn over prairie soil that has been forming for thousand of years. In one day, I can expose 100 acres of soil to the elements. Then I can watch the "tolerable" rate of five tons of topsoil (or more) disappear downstream from one acre of land in the next year.[3]

In many ways, the change from a natural system to an agricultural system is a shrinking down. I was raised on what was once a mixed-grass prairie—a prairie that was a mix of species from both the tallgrass prairie to the east and the shortgrass prairie to the west. Prairies that contained hundreds of species of grasses and forbs (broad-leaved, nonwoody plants) have been reduced to fields of one kind of grass, such as corn or wheat, or one kind of forb, cotton. Beneficial insects that once kept destructive ones in check are lost with the native plants they depended upon, and the boll weevil and a few other pests rule the field. What's ironic and sad about this kind of destruction is that not only is an agroecosystem like this an absolute failure in ecological terms, it is also a failure in human and economic terms. Often these monoculture crops would be unprofitable were it not for government subsidies. Long-time advocate of sustainable agriculture J. Patrick Madden has spoken of the two opposing philosophies or paradigms in agriculture today. One he calls the "harmony with nature" paradigm. It holds, basically, that farmers should try to understand and work with natural processes rather than try to overcome them with technology and power. It is a long-term approach.

He calls the opposing view the "domination of nature" paradigm. It advocates the exploitation of natural resources to produce the maximum output of high-value commodities. This view emphasizes the short term. Madden goes on to say, "This world view separates humans from Nature, and relies on synthetic chemical pesticides and fertilizers with little or no apparent concern for possible detrimental effects on human health, ecological integrity, or long-term productivity for future generations."[4]

I think it is clear that the domination paradigm has not protected the environment and has not helped most farmers, either. Otherwise, we would not have had such economic misery on the farm the past several decades. It seems rather to have benefited major food corporations (such as ADM and the like) and makers of agricultural inputs. It's time to try a new approach. The challenge: figuring out how to incorporate nature's cycles into the farm system in such a way that a profit can be made both in the long term and the short term; ignoring either time frame is unsustainable.

At the farm level, the next three steps go a long way toward making a farm or agroecosystem like a natural ecosystem. I think they can help farmers save money, too. They are: select plants and animals adapted to the environment; encourage biodiversity; and manage pests with minimal environmental impact.

Although often the techniques used to implement these goals require more skill and knowledge from the farmer than what I refer to as "recipe" farming, there is a payoff: a farm that is healthier, more stable and self-sustaining, and, thus, more profitable over the long term. For society, meeting these goals means less pollution, fewer dollars spent on victims of pesticide poisoning, more and cleaner water, fewer endangered species, more genetic diversity of plants and animals, which is a kind of insurance for the future, and, ultimately, a safer, more stable food supply.

BATTLING THE HILL

About the time the Kerr Foundation became the Kerr Center for Sustainable Agriculture, I began renting eighty acres of land adjacent to the 160 acres I had bought in 1977. This parcel of land was, like the home place, a long, narrow strip of pasture next to the road and a long, low, tree-covered hill behind it. Renting this land allowed me to expand my herd of mother cows; the owner assured me that I would have first chance at buying it should it come up for sale in the future. As part of the rental agreement, he stipulated that the pasture be kept brush-hogged, as he had always done. For those who are unfamiliar with a brush hog, it is an implement much like a gigantic lawn mower that is dragged behind a tractor. The heavy, whirling steel blade mows down about any plant it comes in contact with, including, as the name implies, any brush—saplings, vines, and small bushes.

The first years of transition at the Kerr Center, as I have described earlier, were rough, somewhat akin to being the brush beneath the brush hog. Everyone, including me, was struggling to really understand what having a sustainable agriculture meant. Life was moving on at a quick pace. We were experimenting with raising sorghum, with draft horses, with using fewer chemicals, and with sheep and goats for weed control, to name just a few projects. At the same time, we tried to maintain contact with the farmers and ranchers in Le Flore County we were charged with serving—farmers who were alternately hostile toward, confused by, and interested in our work.

Even with some small victories—well-attended field days and a cover story in the state magazine, *Oklahoma Today*—it was rough. Equally rough, I discovered, was brush-hogging my rented land. Although most of it was open and easy to mow, there were three acres on the side of the hill that were rocky and eroded. The owner had cleared it of oaks and attempted to keep it clear. I say attempted because it wasn't easy. Every year, I literally donned a hard hat and went out to do battle with those three acres. I dreaded it. It was both dangerous and hard on my equipment. Every year, I thought if I could just keep those trees from growing, grass would take hold; and every year, when I came back, the grass was thin to nonexistent and a fresh crop of cedar, oak, and pine had sprung up in the crumbly shale. This went on for the five years I rented the land.

When I got the chance to buy this land, I did. But the funny thing was, I did not abandon the task I had come to dread. I went out the next spring and vowed to keep fighting. I would half-jokingly tell my wife to check on me after awhile to make sure I hadn't been killed by one of the rocks that were constantly being thrown up and pelting me in the back and in the head.

At this point, I should have known better. After all, it was just three acres out of 240. Even if it had grown a lush crop of grass, it would have fed maybe one cow. But I still didn't want to stop—though I probably would have said that I didn't want to "give up." Because, without realizing it, I had fallen into the habit of thinking that pervades agriculture. It is stubbornness, a need-to-overcome-nature mentality that came originally, I think, from settling the frontier. As Wheeler McMillen puts it in his excellent book, *Feeding Multitudes,* "[the pioneer farmer] had not time for idleness and a minimum of time for rest. The more trees he downed and burned, the more corn or wheat he could plant. He might, and did, often acquire more land. He saw a future and toiled to attain it. . . ."[5]

I come from such a tradition, and I was proud of its work ethic. My father had worked his way up from sharecropping to owning a small farm, and I now owned more land than my father had dreamed possible. Like my father before me, I wanted my land to be productive. For five more years I kept battling that hill until, finally, one fine spring day, as I sat astride my tractor, engine roaring, rocks exploding all around me like mortars, I was hit with a revelation. "Why am I doing this?" I asked myself. "Every year, nothing changes. Next year the trees will be back."

So I quit. The job was only half done, but I turned the tractor around anyway and went back to the house. I felt a little like Paul on the road to Damascus; I was a changed man. The hilly three acres wanted to grow trees and were meant to grow trees. I would let them grow.

At the time, I knew full well that one of the basic tenets of a sustainable agriculture was to adapt crops and livestock to the ecosystem. I knew that it was folly to believe that one could just force an ecosystem into a desired shape, just as it was folly to assume that the *Titanic* could sail unscathed through the ice fields of the North Atlantic. By this time, I had devised my steps to a sustainable agriculture and had even given speeches on them. This goal of adapting crops and livestock to the environment appeared in print in a number of our publications. But knowing something and applying it to your own situation are two separate things. The experience was sobering and helped me to understand better the many barriers, including the psychological barriers, to adopting a sustainable agriculture. It also caused me to look more closely at my surroundings.

I felt a little like Paul on the road to Damascus; I was a changed man.

Know Your Natural Environment

The Kerr Center ranch is in Le Flore County; I live just over the county line in Latimer County. Both counties are divided into two ecoregions—in the northern part of both you find the Arkansas River Valley region; in the south, the Ouachita Mountain region. Both my farm and the ranch are on the southern edge of the Arkansas River Valley region—sandwiched between the Oklahoma Ozarks to the north and the long east-west ridges of the Ouachita Mountains to the south. In its natural state, this area features bottomland hardwood forests that occur along streams, and dry forests of post oak, blackjack oak, and hickories in the rugged hills. On the ridge tops are shortleaf pine savannas. Scattered in the valleys between the upland and bottomland forests are tallgrass prairie communities composed of native grasses and a wide variety of wildflowers. Indeed, in the spring especially the area is lovely as wildflowers crowd the roadsides and meadows. Historically, fire kept brush from reclaiming these prairies.[6]

Before intensive settlement, the vegetation on my farm was at least similar to the way it is now, with scrub oaks on the rocky hill and grasses on the flat below. The prairie grasses have been replaced by domesticated grasses. The open areas are suitable for grazing cattle.

However, some people are never satisfied with what is appropriate. In the 1970s, when credit was readily available to would-be farmers and ranchers, some of the hills like my own were cleared with bulldozers and grass was planted. It didn't work; most have reverted to trees, but not before erosion took its toll. Here and there, one can still see erosion scars on a bare hillside.

Several years ago, an environmental quiz made the rounds. As I recall, one was asked to name five trees, birds, mammals, etc., native to your area, and identify where your drinking water came from. A lot of people failed, of course, having a better grip on the 7-Elevens than they did on the species of wildflowers in the community. And that was the point: to make the test-takers more aware of their natural surroundings. I can't help but think that farmers, more than anyone, would benefit from such a quiz. If farmers want to make their farms sustainable, they should know the natural environment they are working with.

This is important for all kinds of farmers—from the so-called "hobby farmers," with forty acres or so, to the full-time farmer raising soybeans on 1,000 acres in the Arkansas River bottom. Both of these exist in Le Flore County and the combination is becoming common in rural areas all over the country. For new farmers, especially those with an urban/suburban background, knowing the natural environment will help them decide how best to use their farms, and avoid costly mistakes.

Experienced farmers probably already have a rudimentary knowledge of the natural biota on the farm. They have probably picked plums or blackberries in the summer, and perhaps even gathered a handful of wildflowers. Farmers may not know their scientific names, but they know what season they arrive in, which are common, and which are not. But, like me during my battle with the hill, they may not have been listening to what these plants have been trying to tell them about their farms.

One goal of sustainable agriculture is to have an agroecosystem that maintains itself in a state of optimum health. Matching the crop to the environment is crucial if a farm is to be sustainable. The farm as an ecosystem benefits from having plants and animals that are adapted to it. Why? Plants and animals that fit with their environment are more self-sustaining. They don't need as much human intervention to keep them going, as much fertilizer and pesticides, or in the case of livestock, medication or supplemental feed. Sometimes I think of these inputs as a kind of expensive medicine, an intervention to bring the patient, in this case, the farm, back to some kind of health.

Another goal of sustainable agriculture is for the farmer to make a profit. Some livestock or plants may be able to grow on a farm, but not well enough to make money. Therefore, species must not only be biologically adapted, but also economically adapted to current markets. Farmers must know the range of prices and production costs over which the plant or animal can profitably produce. Even if it is technically feasible to grow a crop, it doesn't mean it is economically feasible.

THE RIGHT COW FOR THE RIGHT GRASS

When Oklahoma senator and oil millionaire Robert S. Kerr came to Le Flore County in the 1950s he bought a lot of land—tens of thousands of acres of mountain, river bottom, and in between. The Kerr Angus Ranch was established along the Poteau River. The senator-turned-rancher had cleared large areas of bottomland that were lush with oak, elm, and hackberry. This was a forest of tall trees (100 feet high) with a tiered understory of smaller trees and bushes. A thick layer of leaves and other litter covered the heavily shaded ground.[7]

You can't raise cows in such a forest. The alteration was drastic—bulldozers were brought in, the wet bottomland terraced for drainage and planted it in what are called "improved grasses," pasture grasses such as Bermuda and tall fescue not native to the area, but adapted to the climate and productive, especially if fertilized. The more upland prairie areas of the ranch, land much like that of my farm, were also turned to pasture.

The Kerr Ranch served as an agricultural example to the community. Until the Great Depression, the county was a leading cotton producer. By the time Kerr came to the area, the agricultural economy had been shifting away from row crops toward cattle. Ranchers in the area emulated the changes Kerr made on his land as well as his choice of cattle and grasses.

After the senator died on New Year's Day 1963, the foundation that bears his name continued to manage about 4,000 acres of the original ranch, and specialized in advising area ranchers and farmers on management of their land. The management was conventional, but conservation oriented. Kerr had been a conservationist in the sense that he had recognized the dangers of soil erosion and supported efforts to slow it. He had planted pine trees on the scarred slopes of nearby Poteau Mountain, later writing in his book *Land, Wood, and Water,* "I cannot describe the joy of planting under the sun and the quickly moving clouds. It gives new faith in tomorrow."[8]

Kerr died less than a year after Rachel Carson's *Silent Spring* was published, and well before the environmental movement really got started. Some of his actions, such as the clearing of forests and draining of wetlands along the Poteau River, would clearly be challenged by conservationists today.

The changes made by the senator in establishing his dream ranch have proved to be both sustainable and unsustainable. The problems on the ranch are the kinds of problems that farmers and ranchers encounter all over the world as they replace natural ecosystems with agroecosystems. What we have done to make the ranch more sustainable provides an interesting illustration of just what kinds of things can be done toward that end and, hopefully, shows those unfamiliar with farming just what farmers and ranchers are up against in attempting to change toward more sustainable operations.

The Kerr Center's goal has been to make the ranch more self-sustaining in the way that a natural ecosystem is self-sustaining. As former Kerr Ranch manager Will Lathrop has written: "The sustainability of cattle producers and the beef cattle industry hinges on the development of more energy efficient production systems that rely less on fossil fuel derivatives (fertilizer, herbicides, pesticides, supplemental feeds, etc.). . . ."[9]

Much of the corn grown in the United States is fed to livestock. Feeding cattle grain is a wasteful practice. Corn is heavily fertilized and sprayed with pesticides made from fossil fuels, and therefore fossil-fuel intensive. Add to that the fact that corn makes beef fattier (and, alas, gives it a taste and texture Americans have grown to love), causing health problems in people who eat too much of it, and the argument is persuasive that we should cut our consumption of beef.

On the other side of the debate is the fact that there are some areas of the United States, such as many parts of Oklahoma, that are best suited for raising cattle. The climate is too dry, or the soil is too thin or highly erodible, or there is some other reason that would make breaking out the land for crops a bad idea. In such areas, the buffalo, another hoofed grasseater, once roamed. And make no mistake about it, land with a grass cover is less erodible than a bare field.

Animals are part of natural ecosystems. As Sir Albert Howard pointed out, nature farms with livestock. Livestock return nutrients to the soil in the form of their manure; they add diversity to the farm and add value to farm products (i.e., beef is more valuable than grass). Add to that the fact that as we seem unwilling to give up our love affair with the hamburger, it seems appropriate to keep producing beef, but in a radically different way.

The solution, it seems clear, is raising forage-finished beef. Ruminants such as cattle can spend their lives making protein out of grass on land that can't be farmed. I predict that eventually feedlots, where cattle are fattened with grain, will be gone.

The Kerr Center's goal has been to make the ranch more self-sustaining in the way that a natural ecosystem is self-sustaining.

So we began a two-pronged experiment. One was to stop fertilizing and applying pesticides to the pastures, thereby allowing the plants there to adapt to natural conditions. The other was to adapt our cattle to the plants.

We started from the ground up. One of our first actions was to do a botanical inventory of the ranch's pastures, woodlands, and riparian areas. As it turned out, the cows shared the spread with more than 500 plant species, including such obscure plants as pencil flower and porcupine grass, as well as the familiar oak and pine. As new management approaches were tried, the survey could be used to assess their impacts.

The next step was to stop applying fertilizer and herbicides. We wanted adapted forage in our pastures that could thrive without a lot of inputs. In order to take full advantage of cow manure as a natural fertilizer, we started managing the cattle herds more closely. We instituted a rotational grazing system. Instead of letting the cattle wander where they would, we controlled their access to pastures, moving them around in order to give pastures time to "rest" and grow adequately between visits by the cattle herd. This mimics the action of the buffalo herds that used to wander Oklahoma. Since we instituted these changes, the plant composition in the pastures has diversified, and plants that thrive without fertilizer, such as clover and other legumes, have increased.

Breeding Adapted Cattle

One of our goals at the ranch for the past ten years has been to develop a cattle breed that was adapted to grass; in other words, a breed that would flesh out (gain weight) easily on grass, thus requiring a minimum of grain to fatten it at the feedlot. Many breeds of cattle require a long feeding time in the feedlot. We wanted a breed that, although eating mostly grass would, nonetheless, have a carcass that would be tender, have good marbling (the threads of fat found in the meat), and be of consistent grade or quality, so that the rancher would get a decent price.

This ideal breed would thrive not just on grass, but also on our grass, in our pastures. We also wanted a breed that could take the heat and humidity of southeastern Oklahoma, and also had traits that every rancher likes: a good disposition, good mothering ability, and easy calving, among others.

What we started with in 1987 was a herd of purebred commercial Angus cows and a herd of cows that were part Angus and part Brahman. The Angus was Senator Kerr's favorite and is known for the exceptional quality of its beef. While he was alive, the ranch boasted 1,000 head of Registered Angus cattle, led by the prodigious bull Hyland Marshall, who sired 7,000 calves over his lifetime.

The Angus, however, was not particularly well adapted to the ranch. For one thing, the Angus, a Scottish breed, does not like the heat. Being solid black in color, an Angus cow is a walking solar collector, which is not a good thing on those 100-plus degree summer days in Oklahoma. It also did not thrive on one of the predominant grasses growing on the ranch—the tall fescue variety Kentucky 31.

The Kentucky 31 tall fescue was another import to the ranch. Senator Kerr planted it in his newly cleared bottomland pastures in the late 1950s, among the first pastures planted with it in eastern Oklahoma. It later was enthusiastically promoted to area ranchers by the Kerr Foundation. Fescue is a cool-season grass—it is green during the winter when nothing else is and provides a valuable source of protein for cattle. It grows well in wet soil, such as the Poteau River bottomland, and so grew well on the ranch.

It seemed like the perfect grass. It wasn't until later that it was discovered that often the Kentucky 31 variety is infected with an endophyte fungus. The fungus-infected fescue is actually good in many ways. It is easy to establish, tolerant of drought, insects, and diseases, and will persist after being repeatedly grazed to the ground. Without the fungus, the fescue isn't nearly as tough.[10]

However, the fungus-infected fescue can cause a number of strange health problems in cattle. The fungus is a basal constrictor, meaning it constricts the cow's blood supply. Extremities are affected first—tails can fall off and, in extreme cases, hooves can rot in a condition called fescue foot. It causes cattle to run a low-grade fever—which explains why some affected cattle stand around in ponds to cool off, even in the winter. Not surprisingly, cattle poisoned by the fungus do not gain weight as well as other cows and look unhealthy when taken to market.

The connection between the fescue and these strange symptoms was not discovered for more than twenty years after it was planted on the Kerr Ranch and on many other ranches throughout southern Oklahoma and the southeastern United States. At the ranch, it turned out to be a worst-case scenario. Not only was the place knee deep in endophyte-infected fescue, but Angus cattle, in particular, seemed particularly vulnerable to the effects of the fungus. It was the kind of situation that makes one humble.

Facing a clear case of nonadaptation, there were two choices: eliminate the fescue or dilute the effect of the fungus by interplanting clovers with it. We chose this latter, moderate approach. After all, the fescue, while not native, was well adapted to the soil and climate. Getting rid of it completely would be costly and time consuming and would leave the cattle with no winter forage.

We also decided to try to find a cow better adapted to our forage, including the endophyte-infected fescue, a ten-year quest that is not yet complete.

Cross-breeding cattle for desirable characteristics is a slow process. A cow's gestation period is nine months; calves are not weaned until they are 205 to 240 days old. It can take ten years to change the genetic makeup of a cattle herd. Ranch manager Lathrop experimented with crossing Angus with various breeds known for their tolerance to heat and, hopefully, tall fescue. He worked with Gelbvieh, a German breed noted for fast growth, quality carcass, and good maternal care. He also worked with Brahman (known locally as Brahma), the humped cattle of India, which, as might be expected, don't mind the heat. I have observed Brahmas out grazing in the sun while cows of an English breed take to the shade.

Despite a number of other good traits such as tolerance of the endophyte-infected fescue, good mothering, and all-around toughness, the disposition and carcass quality of the Brahma left something to be desired. Also, beginning in the 1980s, there was a gradual marketing problem with them. Prices for animals that looked as if they had some Brahma blood in them (humped, large floppy ears) were discounted. For our part of the country, we sorely needed an animal that had the Brahma's strong points, without its marketing minuses.

In 1991, Lathrop started crossing Senepol cattle with Angus and Angus-crosses. Like the Brahma, the Senepol is also tolerant of heat. The Senepol is a good example of how far a cow can come. The breed is from the Caribbean island of St. Croix—a cross between an N'Dama cow from Senegal, West Africa, and the Red Poll, an English breed. Lathrop has found that the crossing of Senepol with Angus and Gelbvieh-cross cows "has produced a biological type that is superior to Angus when the cattle are required to utilize infected fescue and superior to Brahman in disposition, age of maturity, and marketability. The Senepol replacements gradually formed a new heat-tolerant herd capable of utilizing tall fescue pastures."[11]

For the Concerned Consumer:
What You Can Do to Encourage the Use of Adapted Crops and Animals

1. Grow a garden and do your own experiments with adapted varieties.
2. Know your environment—what foods grow wild in your area? What crops seem to be especially abundant and well adapted?

(continued)

(continued)

3. Buy locally produced food in farm stores, farmers' markets, and groceries. (Although such venues do not always offer adapted crops, they often do. For example, Oklahoma has many pecan farms and an abundance of wild, native pecans. For Oklahomans, the clear choice would be local pecans over California walnuts to flavor baked goods. Consumers in all parts of the country can make such choices daily.)
4. Ask your grocer to buy locally grown food.
5. Encourage institutions such as schools and hospitals to buy locally grown food.

Along with creating a fescue-tolerant cow, we have tried to select bulls that seem to gain the most weight just eating grass. At intervals during the growing season, our young bulls are weighed to see how they are faring on our forage. The ones who gain best we keep or sell as forage performers. In this way, we hope to have a herd that is not only adapted to the conditions on our ranch, but will be economically adapted as well: cheaper to produce, more conserving of natural resources, and healthier for the consumer as well.

Our goal is to produce an animal that will grade "choice" on grass. However, the current market still favors more fat than we can get on our animals by feeding them just grass. We have, however, cut substantially the amount of time our cattle are fed grain, down from four to five months to about three months. We believe we are on the right trail with the Senepol/Angus cross, and are looking to develop this breed until we have an animal that is about 5/8 Senepol and 3/8 Angus—a good-looking red cow that is well adapted to the area.

THINK LOCAL

"All politics are local." Any time I watch election returns on television, I hear someone say this at least once. It is a truism. I would like to modify this saying to "All farming is local."

This truth is often overwhelmed by hype and greed. A case in point: anyone who has perused a rural newspaper has seen advertisements for varieties of grass that can grow a foot overnight. Gullible farmers sprig their pastures with the stuff only to find that after a year or two of stupendous performance it is killed out by a hard winter. The grass, it turns out, was not really adapted to Oklahoma, but to a more southern climate, something not mentioned in the ads.

It happens with livestock, too. I first learned about the perils of non-adaptation when I was in high school. I showed pigs at the county fair, and I wanted to win. So I ordered two prize boars from an Indiana supplier. With high hopes, I spent all my money on buying the animals and having them shipped to Cold Springs. A few months later, my hopes had been completely dashed. These Indiana pigs did not like the blistering hot Kiowa County summers and simply were not interested in breeding; in fact, they were not interested in anything but finding a cool mud hole.

The fact is, selling seed and bloodstock is a high-profit enterprise, and sometimes marketers overpromote new varieties that haven't been tested by time or the environment. This hype feeds the desperate farmer's dream of something big to lift him out of the hole of debt he is in.

Such was the case described in Victor Davis Hanson's book, *Fields Without Dreams,* in which he describes his extended family's attempts to make a living on their raisin and fruit farm in California. Hanson fell victim to exaggerated claims for new varieties of grapes and plums; alluringly big and beautiful, they both failed miserably on his farm. As Hanson put it, "Born out of the laboratory, the strange concoctions of plant alchemists, neither species could be grown under the real conditions of the vineyard or orchard. They were fruits whose dazzling harvest required either a secret and unknown process of pollination or a permanent fog of toxic chemicals." That these varieties had been put on the market at all, Hanson chalks up to "stealth and connivance of food brokers . . . [that] brought the grape into the daylight world of commercial production."[12]

This triumph of looks over reality is, in fact, all too common in farming today. One sees it in the show ring at local Future Farmers of America or 4-H events. A pig with such a big ham that it can't reproduce or a steer that wouldn't produce a volume of high-priced cuts are selected as champions. Looks win out over utility, and our kids are learning the wrong lesson. This is true in gardening, too. Why else would, as in one case, cherry tomatoes adapted to local conditions and resistant to a number of pests and pathogens—continuous producers which reseed themselves each year—be frequently passed over in favor of big tomatoes produced from special hybrids?[13]

What farmers need today are plants that are resistant to local insects and tolerant of local weather conditions, such as droughty summers, instead of hybrid plants selected only for great yields under ideal conditions. These hybrids are the norm today. Although they can outyield older varieties, given the right conditions, they are not necessarily adapted to local conditions. Farmers used to save seed from their best crops; now these local varieties are largely out of circulation. I think farmers need such varieties today; even if lower yielding, they would be more dependable and require fewer inputs of fertilizer, water, or pesticides, and therefore offset lower production through lower costs.

What farmers need today are plants that are resistant to local insects and tolerant of local weather conditions, such as droughty summers, instead of hybrid plants selected only for great yields under ideal conditions.

Adapting crops and animals to the farm environment is more than just an economic benefit to the farmer. It is a key component of true food security. The taxpayer is already footing the bill for disaster payments, doled out to farmers participating in government programs. While they have some merit, their cost to the taxpayer is high. The use of crops adapted to the local area can ease the strain on the taxpayer and on consumers who pay higher prices when crops fail.

Of course, natural disasters are often just that. But the risk can be lowered. Of great concern to me and others in sustainable agriculture is the risk of disease devastating major agricultural crops that are narrowly bred for production under the assumption that there will always be a chemical solution for disease or insect attack. Another concern: plants being bred for conventional systems that depend on synthetic chemicals instead of natural fertility. Modern farms are chemically dependent, with fields requiring an annual fix to keep going. Now we have genetically engineered crops, such as 'Roundup Ready Soybeans,' which demand that farmers use a specific chemical on their field, in this case the herbicide Roundup. If one doesn't need or use Roundup, the premium price paid for these beans would have been wasted. What will happen when prices of inputs are simply not affordable relative to prices received by farmers for their crops?

Adapting crops and animals to the farm environment is more than just an economic benefit to the farmer. It is a key component of true food security.

It makes sense, then, to raise crops and livestock that are adapted to the local environment as much as possible. I believe that means looking at the natural environment for clues as to what works best. Some Oklahoma farmers are doing just that. Although many Oklahoma ranchers continue to spray and brush-hog every blackberry vine they see, considering them weeds, a few have taken to raising blackberries, which thrive here, and selling them for premium prices. One entrepreneur in western Oklahoma has begun selling seed packets of "weeds" as Oklahoma wildflowers.

Checklist for Farmers: How to Select Plants and Animals Adapted to the Environment

1. Know the plant life. Survey the existing vegetation on your place. Pay particular notice to plants in the fence rows or in undisturbed areas for clues as to what is native or grows well.
2. Know your climate, including rainfall patterns and average first and last frost dates. Identify microclimates created by shade, slope, streams, etc.
3. Know your soil. County soil survey maps are available from the Natural Resources Conservation Service (NRCS) offices. They include detailed information, section by section, on the soils of your county. Included is information about depth, drainage, recommended uses, and information on adapted plants.
4. Talk to your neighbors. A person who has worked the land has plenty of stories of successes and failures. Learn from them.
5. Test your soil.
6. Find out the cropping history of your place.
7. Go to see test plots. Your state land-grant university often has test plots scattered around the state. When visiting these plots, ask these questions: What is the soil like? What inputs were used? Are the crops being grown using the "Cadillac treatment"—i.e., with maximum inputs to maximize production? How long has the crop been tested? How do climate and other conditions compare to your own? Are they similar enough for you to expect similar results?
8. Plant your own test plot.
9. Experiment with older, open-pollinated crops that do well without chemical inputs.

(continued)

(continued)

10. Raise livestock that gain well on native forages or forages that do not require expensive inputs.
11. Raise livestock adapted to your climate.
12. Raise crops adapted to your local conditions.

Perhaps if this philosophy had been in place forty years ago, when Senator Kerr was carving out his ranch, he would have left more of the native grasses in place, rather than importing a grass that has proved to be a mixed blessing. It seems that we're still learning. I know I am; the three acres I assaulted with my brush-hog for ten years are rebounding with trees. It is a confirmation of the truth that it is trees that grow best there and a definite sign of the resiliency of nature. Building this resiliency into agricultural systems is a worthy goal.

If I wanted to, I could eventually harvest some trees for firewood or fence posts. Then again, I might let them be. After all, what a tree does as a tree benefits me—with the oxygen it produces, the carbon monoxide it absorbs from the highway near my house, the soil it holds in place, and the rocks its roots are slowly crumbling into good soil. A growing tree reminds me once again that nature has a perfect intelligence.

Chapter 6

All Creatures Great and Small: Step 5—Encourage Biodiversity

> It is often said that variety is the spice of life. No intelligent investor confines his money to one or two shares. No one can sit stably and comfortably on a chair with two legs. No one remains fully healthy on a restricted diet. These facts are obvious, but the larger analogy that a varied base is vital for human existence fails to achieve recognition.
>
> Prince Bernhard of the Netherlands
> Founder President of the World Wildlife Fund[1]

If you look in the dictionary for the origin of the word *farm,* you will find that it came originally from the Latin word *firmus,* which is also the Latin origin of the word *firm.* Firm: unyielding to pressure, solid. Is a farm, too, solid and unyielding to pressure? The Romans thought so, and you can see why: in those times, farmers labored hard with their hoes and small plows to break open the body of the earth and grow a crop. The earth was *terra firma*—solid ground, unyielding, unchanging. It is an ancient concept, even a comforting one.

These days, we don't have that comfort; we know that neither Mother Earth nor farms are unchanging. We have realized that agroecosystems, like other ecosystems, do indeed change under pressure—and we are very good at applying pressure. As we enter the twenty-first century, we bring with us a fantastic array of tools and machines that enable us to literally move mountains. It takes today's farmer just three labor hours to produce 100 bushels of wheat from three acres of land. To do this, a tractor big enough to pull a 35-foot-wide sweep disk, a 30-foot drill, and a 25-foot self-propelled combine are used. The Romans would be amazed.

An ecosystem is like a mobile, where any piece added or subtracted from it affects its balance. Even laying something on one of the pieces can cause the equilibrium to shift; and, of course, the bigger the change, the more the mobile is thrown out of its former balance.

A case in point: As a child, I observed closely the life cycles of the rabbit and the coyote, two common animals of farm country. As rabbit populations increased, my father would predict that in a couple of years we would see an increase in the number of coyotes who fed on rabbits and rodents. Sure enough, it happened. It was a good lesson in how a natural system can regulate itself and return to balance.

Then coyote hunters, rich city folk, would lease helicopters to "hunt" coyotes. It was an unequal match—the silent, running coyote versus the roaring flying machine carrying shouting men with high-powered rifles. The coyotes always lost. The hunters would hang them on the fence posts as trophies, fifteen or twenty in a row. Once I counted thirty-seven. I thought then, and still do, that this was surely an abomination to God.

After such episodes, the rabbit population would multiply, unchecked by the coyotes for awhile. They would start eating the young cotton plants, gardens, and other food crops. The coyotes, if given a chance, would eventually rally to take advantage of this bumper crop of rabbits, but until then we bore the consequences of interfering with a natural cycle and throwing the farm out of healthy balance.

The balance of rabbits and coyotes was also a favorite 4-H club lesson that illustrated for farm kids the concept of the balance of nature. To maintain that balance, it is important to encourage a diversity of life in an agroecosystem. Biodiversity is the word used today, short for biological diversity, which is the variety that exists among organisms and their environments. It refers to diversity on several levels, each of which is rich and complex.

The first level is diversity of habitats or ecosystems. A farm can contain a variety of habitats: ponds, streams, wetlands, woods, strips of scrub, thickets, orchards, meadows, pastures, and, of course, fields of crops. A farm that leaves some habitat for wildlife would be more diverse than a farm that is plowed from fence row to fence row. One habitat that sometimes goes unrecognized is the soil, which is a world of its own, with plants and animals living and dying within its depths.

Next is the diversity of species within each habitat. A biodiverse farm has not only a diversity of wild species of plants and animals, but also a diversity of crops and livestock. Thus, a farm that raised cattle, sheep, and emus as well as corn, tomatoes, and alfalfa would be more biologically diverse than a farm that raised only corn and wheat.

A habitat that needs a diversity of species to function well, as mentioned, is the soil. Although we cannot see them, the tens of thousands of microbial species in the soil and on plants are essential for decomposing crop residues and cycling nutrients. A farm that preserves the health of the microbial life in its soil would also be promoting biodiversity. So would a farm that allows habitat for insects, which are needed to pollinate crops and protect them from pests.

The next level is the diversity of genetic material within a species. This is perhaps the easiest to overlook. To plant varieties of crops that are genetically close or identical may leave them open to devastation by disease, insects, or extreme weather conditions.

A farm that is diverse on each level has a better chance of being sustainable, both economically and ecologically. Sir Albert Howard addressed biodiversity when he said: *Mother earth never attempts to farm without live stock; she always raises mixed crops. . . .*

Biodiversity and Diversification

Industrial agriculture is mostly silent on the subject of biodiversity. I don't believe I heard the word until ten years ago when I began learning about sustainable agriculture. Biodiversity is not part of the industrial agriculture equation. The emphasis is on production, and this often means conversion of every square foot on the farm to growing space for one or two crops. Often these fields of crops are not diverse genetically, and instead are planted in one or two hybrid varieties, which may also be quite similar genetically. Industrial agriculture has also promoted the use of pesticides, which often kills many species besides the target.

Biodiversity is a word from the biological sciences, not a word most farmers are familiar with or bandy about much. However, I believe it is a word that we will become more familiar with in the coming years because it is so important to farmers, as well as everyone else. Biodiversity is key to human survival.

On the other hand, the word *diversified* is more familiar to farmers. A diversified farm has a variety of crops and livestock. It implies a certain measure of biodiversity. We rented five farms at varying times as I was growing up in Kiowa County. Our Cold Springs farms, taken all together, were diversified because we grew a variety of crops—cotton, wheat, barley, maize, alfalfa, and maintained pastures of buffalo grass with mesquite. We also raised a variety of animals—pigs, chickens, and cattle. A diversity of enterprises on a farm was still the norm when I was growing up. As it turns out, it wouldn't be for much longer. Although diversification is still touted as a good thing by industrial agriculture, in reality, farmers have moved away from it toward specializing in one or two crops.

A diversified farm has a variety of crops and livestock. It implies a certain measure of biodiversity.

As an agricultural economist with an interest in making farms sustainable, I believe in diversification of farm enterprises, the more the better. My father had good reasons to mix up his crops; a diversified farm gains economic resiliency and stability. If the price of one crop is down, or there is a crop failure, growing another crop or livestock can help provide a safety net.

Although overall my family grew a diversity of crops, on some fields we grew just one crop over and over again. For example, the farm on which we grew cotton would have been better for us financially if we had diversified—if we had rotated crops—instead of only growing cotton each year. The soil would have benefited, too. But because we did not own the land, and because of government allotment programs, we could not do it. We were sharecroppers to a big landowner. He didn't like farming cotton, so he transferred all of his cotton allotments to this single farm, which he rented to us.

An allotment is the number of acres a producer is allowed to grow of a particular crop under a government program. In order to keep the right to grow cotton, a producer had to consistently grow cotton—in other words, use it or lose it. So my father didn't make this farm a monoculture by choice; we had to do it to get land to farm.

Monoculture or monocropping is the cultivation of the same one or two crops from year to year. USDA programs have encouraged monoculture on individual farms, which has led to large regions of monoculture—a drive through the Oklahoma panhandle in June when the wheat is ripening or Indiana in July when the corn is growing confirms this. To opt for putting land into crops that were not covered by price supports or that would not be covered by disaster payments was not an option for many farmers. This situation made diversification an economic liability.

So farmers have continued to overproduce certain commodities, which leads to low prices, which leads to bankruptcy and the consolidation of farms, followed by more production in an attempt to make up for lower prices by growing even more. It's a vicious circle that continues to this day, producing recurring farm crises, the latest of which began in the late 1990s.

The upshot is that now most farms are not truly diversified. And although the allotment system was phased out by the Freedom to Farm Bill of 1996, farmers have found that decades of farm programs have created a production, processing, and marketing system based on a few crops, making it difficult to diversify.[2] Farmers continue to raise standard commodity crops such as corn, soybeans, and wheat simply because they can haul these crops into town and sell them, albeit at prices that are too often below the cost of production. If you grow a crop not commonly grown in your area, there is often no way to market it locally.[3] If you want to grow something really unusual, such as an organic crop, you might have to search far afield for a market or even direct market it yourself.

Although miles and miles of waving wheat or rustling corn is picturesque, monoculture is risky from both an economic and ecological point of view. Imagine a farm as a brick wall. A farm that is a monoculture is a wall built of just a few large bricks; a diversified farm is a wall built of a large number of smaller bricks. If something comes along and knocks a brick out of these walls, which is more likely to remain standing? That something could be drought, disease, or insect infestation, as well as low prices.

Monoculture leads to ecological instability, too. Industrial agriculture is modeled on a factory: supply the raw materials—soil, seed, water, chemicals—and out comes the crop, the finished product. It is a simple recipe, a simple model, and at first glance seems to work. In the long run, because the model doesn't match the reality of nature, it begins to break down. The farther agriculture goes toward the factory model and away from a natural ecosystem, the more risky and unstable an enterprise it becomes.

In the Red River Valley of the northern plains of Minnesota, North and South Dakota, and Canada, the outbreak of wheat scab over the past five years is a perfect example of what can happen when a whole region is planted with just a couple of crops, in this case, wheat and barley. The same crops planted year after year in contiguous sections of farmland have made it easy for the fungus to survive and spread. The result is that small grains growers in the region have lost 4.2 billion dollars in income since 1992, mostly because of the scab.[4]

Modeling a farm on a natural ecosystem as much as possible reduces this risk. The reality is that the farm is a part of nature, and natural processes hold sway. In a natural ecosystem there is no monoculture; in a natural ecosystem, biodiversity is the rule.

STEWARDS OF LIFE

Forty years after the coyote kills, preserving the biodiversity of southwestern Oklahoma—either in wildlife or in crops and livestock—is still not a high priority in my home county. In fact, diversification is largely a thing of the past. Many of the pastures that were part of each farm have been plowed for a couple of cash crops—mainly cotton and wheat. Everything is being plowed up—ditches, shelter belts, and even the fence rows.

As a result, biodiversity in the wildlife population has shrunk, too. Besides protecting fields from wind erosion, these strips of uncultivated land were wildlife habitats, providing homes for quail, meadowlarks, armadillos, possums, skunks, field mice, and rabbits. Fence lines near these uncultivated strips were always decorated with scissortail flycatchers (Oklahoma's elegant state bird) and mourning doves. The pastures of native vegetation—buffalo grass and mesquite—provided habitat for myriad creatures, including grasshoppers, diverting them from fields of young tender plants.

Not as much space remains for these creatures, great and small, these days. This is just the latest refrain in an old song: since the landing of the Pilgrims at Plymouth Rock in 1620, more than 500 species, subspecies, and varieties of our nation's plants and animals have become extinct. Even during the 3,000 years of the Ice Age in North America during the Pleistocene epoch the continent lost only about three species per year.[5] The Ice Age extinction rate added up to about ninety species over 3,000 years, compared to the modern record of 500 species in not quite 400 years. And even this rate of extinction is accelerating, as in the rain forests of South America, which are in some cases being turned into farms and ranches. The result: mass extinction. Some estimate one-quarter of all tropical plants may be gone in a few decades. But it is not just in the tropics—3,000 of the 25,000 plant varieties in the United States are threatened as well; 700 varieties are near extinction.[6]

Agriculture cannot take the whole blame for this state of affairs, but farmers bear some responsibility for it historically and currently. Historically, the clearing of land for agriculture has destroyed habitat for wildlife. Sometimes land clearing was preceded by the hunting out of wildlife. The nineteenth-century slaughter of the bison herds on the Great Plains not only destroyed the traditional Native American cultures dependent on them, but opened up the prairies for farming and ranching. Although often not as drastic an environmental change as farming, ranching, too, can be hard on wildlife—overgrazing by livestock can destroy food species upon which wild animals depend.

Currently, industrial agricultural practices are still hurting wildlife. An example: Northern bobwhite quail numbers have been declining since the 1970s, due in part to pesticides and habitat modification.[7]

Half of the land in the United States is agricultural land, controlled by farmers and ranchers. I believe those of us in agriculture have a responsibility to join the fight to preserve the biodiversity of the planet. If we don't, our children and grandchildren will be the poorer. I know that some in agriculture take a dim view of this position. Many in agriculture ask why they should shoulder the financial burden to, say, save the spotted owl. In this, they are no different than the corporation that gripes when new clean air standards are enacted that will cost them money. The difference is, of course, that the large corporation can often pass on to consumers the costs of pollution control devices, whereas the individual farmer cannot. The individual farmer feels embattled—low prices on one side, environmental regulations on the other.

I believe farmers have a valid point, one that should be taken seriously by the government and environmental groups when decisions are being made about what kinds of assistance to provide farmers. At the same time, farmers need to embrace their traditional role as stewards of the land. With our new understanding of the web of life, it may be more fitting to call those of us who live on the land and benefit from its bounty stewards of life. We should do this for two reasons: for the health of the planet we live on and for ourselves. Promoting biodiversity not only helps the greater environment but can also benefit us as farmers.

As an example, look at birds. Birds eat millions of insects and perhaps billions of weed seeds; in essence, they provide us with a free service. Barn owls eat mice and grasshopper sparrows eat insects; they do this largely without our noticing, which is the problem.

However, put to a cost-benefit analysis, birds hold their own. An analysis of the use of granular carbofuran on rapeseed in Canada showed that when the insect control benefit of this pesticide is compared with the value of insect control lost because of the insecticide's killing of songbirds, the use of granular carbofuran actually represented a net loss to the Canadian economy.[8]

With our new understanding of the web of life, it may be more fitting to call those of us who live on the land and benefit from its bounty stewards of life.

Beneficial insects are another example. More than 25 percent of the world's crop production is destroyed by pests annually. More than 90 percent of potential crop insect pests are controlled by natural enemies that live in natural and seminatural areas adjacent to farmlands. The substitution of pesticides for natural pest control services is estimated to cost 54 billion dollars per year.[9] One could extrapolate from this that plowing fence row to fence row and thereby destroying these wild areas is also costing farmers money.

Biologists are beginning to understand the complex interactions in ecosystems, and economists are just beginning to take into account the many benefits and services the natural environment provides for us. In agriculture for the past fifty years, much of our energy has been put into making agroecosystems simpler, not making them more complex. Unfortunately, we often don't know what we have until it's gone; as in the case of the big coyote kills of my childhood, the positive effects of biodiversity are often not noticed until a key animal, plant, or ecosystem is lost. Until, it's sad to say, there is a silent spring.

Ecologists tell us that the removal of a single species can set off a chain reaction in its community. Just one disappearing plant species can cause the disappearance of up to thirty other species of insects, higher animals, and other plants.[10] So it should come as no surprise that changes in agricultural practices that result in the decline of whole communities of plants also have a drastic effect on species that depend on them. According to several studies, factors such as increased field size, a corresponding decrease in fence row and other field-separating land uses, and reductions in the variety of crops produced on farms in the midwestern United States are responsible for dramatic declines in native bird populations, such as prairie chickens, bobwhite quail, and ring-necked pheasants.[11] An Illinois study found that the expansion of row-crop farming and its associated decrease in oat-hay rotations explains over 90 percent of a roughly 50 percent decline in small game taken by hunters.[12]

Small changes in farming practices can yield surprisingly big results. A SARE study found that just leaving fifteen-foot wide field borders could make a dramatic difference in the number of quail that a farm can support. Over a two-year period, agricultural areas with field borders produced 179 quail hatchlings, compared to 37 hatchlings in areas without field borders. Flush counts confirmed that cropped areas with field borders supported about six times more quail than areas without field borders.[13]

. . . economists are just beginning to take into account the many benefits and services the natural environment provides for us.

There are plenty of arguments for maintaining biodiversity on each level. On the ecosystem level, wetlands are a good example. Although how best to preserve wetlands on agricultural land remains controversial, public attitudes toward the value of wetlands are beginning to change. Wetlands—swamps, bogs, mires, and potholes—were once considered wasteland, valuable only when drained. Now we know that wetlands are nature's filters, cleaning pollutants out of water before it filters into underground reservoirs. Wetlands are home to an incredible variety of animals, including large numbers of waterfowl. Economically speaking, hunters, as well as bird watchers, spend dollars in rural areas. Duck hunters, according to one estimate, spend about thirty-five dollars per day. Wetlands are also natural reservoirs, absorbing water in times of heavy rains. When there are record rains, wetlands are crucial. The great Mississippi River flood of 1993, which inundated so much valuable farmland, was made worse by the draining of wetlands for agricultural use.

The Value of Wild Plants

When ecosystems are saved—rainforests, prairies, wetlands, and forests—the plants and animals that live within them are preserved. These ecosystems can provide us with food, medicine, and industrial products. There are an estimated 80,000 species of edible plants on the planet, but according the UN's Food and Agriculture Organization, only eighty plant crops produce 90 percent of our food. There are 250 important edible fruits in tropical rainforests alone, and thousands of potential vegetables. It seems we are not fully utilizing our resources; in fact, in many cases, such as in the tropical rainforests, we are destroying them. It is the worst kind of waste.

Besides providing us with food, plants also perform other valuable services for us. Trees replenish the oxygen in the atmosphere; they moderate the climate; they protect us from the wind. Just when there may be a need for a new plant cannot be predicted, but it happens all the time. During the dust bowl days of the 1930s, the drought-resistant Siberian and Chinese elms, trees native to Asia that had been collected and brought to the United States by USDA plant explorer Frank N. Meyer, were planted in windbreaks that helped conserve millions of tons of soil.[14]

Wild plants can yield important medicines—a lowly fungus gave us penicillin. Of the 150 most common prescription drugs used in the United States, 118 are based on compounds derived from natural sources.[15] It seems that almost monthly plants are being found that do what more expensive synthetic drugs can do, often with fewer side effects. The drug Taxol, derived from yew trees and used to treat breast cancer, is an example of a recent discovery of an important natural drug.

Industrial uses are being found for other wild plants. Two examples are the guayule and the jojoba, plants of the desert southwest. The jojoba yields oil that can be used for a variety of industrial processes. The guayule contains high amounts of natural rubber. Wild plants can also yield important natural pesticides or can harbor beneficial insects that can cut our pesticide bill.[16]

How inconspicuous plants can suddenly take on importance was illustrated in Oklahoma a few years ago by the sudden popularity of a roadside "weed," the purple coneflower or echinacea. Americans had discovered the virtues of this plant, which is much used medicinally in Europe as a booster for the immune system. Echinacea had long been sprayed by road crews, plowed up by farmers, and ignored by everyone except wildflower buffs and Native Americans who knew of its medicinal virtues.

Suddenly, dealers were offering large sums per pound for the plant. No one was growing it, so people were out combing the roadsides and meadows in rural counties, digging up the wild plants. Although such indiscriminate harvesting was negative, the episode highlights how little we know about the value of the many plants and animals on our own land, and how important biodiversity can be.

Echinacea, guayule, and jojoba are plants with names that are unfamiliar and much harder to pronounce than wheat, corn, and cotton. But who knows when a plant with a heretofore unpronounceable name may turn out to be important to the world (and profitable for farmers)?

ENDANGERED AT HOME

When ecologists talk about the importance of maintaining biodiversity, they often mean a diversity of wild animals and plants. Biodiversity on the farm also means raising a variety of crops and livestock.

To carry it a step further, there should be variety within species of crops and livestock. This kind of diversity is known as genetic diversity: the variety of genes that exist in members of a species. Take one species, *Bos taurus,* the common cow. Within that species are hundreds of breeds, from the English Ayrshire to the Italian Tarentaise, each different. Some have horns; some don't. Some are big; others are compact. There are dairy and beef and draft cattle, and those that perform in more than one way. Some thrive in hot weather; others cold. They differ in many ways: body shape, color, disposition, size of calves, mothering ability, resistance to disease, and carcass traits, to name just a few. The genetic variation is formidable. This same diversity holds true among other livestock such as swine, goats, and sheep, as well as in poultry—chickens, turkeys, geese, and ducks.

Writing in the *Permaculture Dryland's Journal,* Lynne Trewe made a simple yet profound statement: "Domestication represents history." In the case of livestock, "older breeds embody centuries of selection for multifunctional traits—traits that have withstood the trials of disease and climactic change."[17] The problem is that these breeds are disappearing.

When people think of protecting endangered species, they again think of wild animals and plants. However, more than 100 breeds of livestock and poultry in America are endangered and may become extinct, according to the American Livestock Breeds Conservancy (ALBC), a nonprofit organization dedicated to the conservation of rare poultry and American livestock "heritage" breeds.[18]

Worldwide, 400 native species of livestock animals are in danger of extinction. In Europe, the situation is even more dire—one-half of European livestock breeds are already extinct, and another third of some 700 breeds (from five livestock species) are near extinction.[19]

According to veterinarian Donald Bixby, executive director of the ALBC, today's food supply is balanced precariously on a single breed of dairy cattle, two types of chickens, three breeds of hogs, and one turkey variety.[20] It is another example of industrial agriculture's narrow emphasis on high yields, whether it is grain, meat, milk, or eggs. Unfortunately, these high-yielding breeds are often dependent upon high-energy inputs of chemicals or feed, and high-tech shelter in order to produce at high levels.[21] Genetic traits that might make them more adaptable to change or resistant to disease are not emphasized in breeding programs and may be lost.

When people think of protecting endangered species, they again think of wild animals and plants. However, more than 100 breeds of livestock and poultry in America are endangered and may become extinct.

In animals, the emphasis on yield means selecting animal breeds that show the most expedient conversion of feed into the most flesh or egg or milk. How far the situation can go is illustrated by the Holstein cow. Over 90 percent of U.S. dairy cows are Holsteins. As a result of this emphasis on the Holstein, registration of other familiar dairy breeds, such as the Guernsey with its delicious golden milk, have declined by half in the last twenty-five years.[22] Once a dual purpose Dutch breed good for both milk and draft work, the Holstein has become, through selective breeding, what some call a milk machine, its other traits de-emphasized. This high productivity, however, depends upon on maximum nutrition, a nonstressful climate, and excellent husbandry.[23]

In pursuit of the Holy Grail of high production, other genetic traits fall by the wayside. The ability to adapt to a local environment is simply not needed, for example, in the high-maintenance world of a confined chicken operation. Chickens of one type are kept in expensive, environmentally controlled houses built to exacting standards. To reach maximum production, the chickens require special feed and intensive medical care.

Being so uniform genetically, these CAFO chickens are uniformly vulnerable to new strains of disease. In 1998, an avian flu swept through the large, crowded, confined chicken operations in Hong Kong; the outbreak was made worse by the lack of genetic diversity among the chickens. This is nothing but monoculture for animals.

Meanwhile, other breeds of chickens, such as the White Wyandotte and the Delaware, that have other worthy traits besides how well they convert feed to meat, are listed in the "critical category" of endangerment by the ALBC (fewer than 1,000 breeding females in the United States, five or fewer primary breeding flocks). Five of the seven chicken breeds on the critical list originated in the United States, including the Delaware and White Wyandotte, who are cold hardy and will lay some eggs in winter.

Another historic American chicken is the scrappy Dominique (known colloquially as "Dominiker"), with coloring that makes it less conspicuous to predators and its heavy plumage ideal for cold weather. Although the coddled, industrially raised chicken does not need these traits, who knows what the future may bring? If we let the Dominique, the oldest breed developed in the United States, disappear, what traits—hardiness, disease resistance, foraging ability—might go with it?

With this in mind, the Kerr Center is helping to preserve Dominique chickens as well as a few breeds of livestock listed as critical by the ALBC through a program at our Overstreet-Kerr Historical farm. The farm was established in the 1870s, which is quite old by Oklahoma standards (statehood in 1907), in what was then the Choctaw Nation.

Some of the breeds we have raised are associated with the Choctaws; one is the Pineywoods cow. There are fewer than 200 North American annual registrations and estimated fewer than 2,000 Pineywoods cattle left in the world. Pineywoods cattle take their name from the pine woods of southern Mississippi. It is a small, hardy breed related to Longhorn cattle and originally brought to the Americas by the Spanish. The Pineywoods is a multipurpose breed, used for meat, milk, and draft. It developed through natural selection and can survive and reproduce under the harsh conditions of the Deep South, including assaults from internal and external parasites, high temperatures and humidity, and low-quality forage. They are quite self-sufficient.

The Choctaw Indians who were removed from their ancestral lands in Mississippi to southeastern Oklahoma raised these cattle, as well as the Choctaw hog. These hogs were probably introduced to the Southeast by the Spanish. Historically, the hog was allowed free range, and survived on roots, small plants, berries, and acorns. They are noted for being protective of each other and their young, and according to Jim Combs, who is in charge of endangered breeds at the Kerr Center, they require very little care. This self-sufficiency is in contrast to today's most common swine breeds, who have been bred largely for fast growth and large litters.

How quickly such animals can disappear is illustrated by the case of the Choctaw horse. As late as the mid-1970s, their population numbered in the hundreds; by 1988, only fifty purebreds could be located by the ALBC. A small and sturdy Spanish-type horse, with a big heart and willingness to please, most Choctaw horses have now been destroyed or crossbred in order to make way for a more appealing show type.

It is not just the genetic diversity of livestock that is declining. The same narrowing and homogenizing is going on for food plants as well.

The gene pool for many crops has been considerably narrowed; modern breeders often work with just a few varieties of hybrids that are superior producers. Contrast this to what is available: There are about 125,000 strains of wheat in seed banks around the world, most of which are locally used varieties of the crop developed by farmers for their own use.[24] Crop biodiversity is still the norm in some parts of the world: Indigenous peoples in the Philippines grow over 200 varieties of sweet potatoes, and Andean farmers still dig over 1,000 varieties of potatoes.[25]

However, the narrowing of genetic possibilities is fast overtaking even remote corners of the globe. The result is that thousands of irreplaceable varieties or local strains of crops have been abandoned or lost. The United Nations' Food and Agricultural Organization has estimated that since the beginning of the twentieth century about 75 percent of the genetic diversity of agricultural crops has been lost.[26] Some examples include China, where nearly 10,000 wheat varieties were cultivated in 1949; by the 1970s, only about 1,000 were in use.[27] In Mexico, too, unique, irreplaceable varieties of maize are gone. Only 20 percent of the maize varieties reported in 1930 are now used,[28] this in a part of the world that is the genetic home of maize. These losses are occurring in fruit crops as well. Since 1900, 85 percent of the known apple varieties have disappeared, as well as several thousand pear varieties.[29]

. . . thousands of irreplaceable varieties or local strains of crops have been abandoned or lost.

The Importance of Diversity

When a local population or variety of a plant becomes extinct, it certainly reduces the genetic richness of a species. Why does this matter so much? Because this "genetic erosion," as some call it, will make it harder to meet change head on. One can think of varieties of crop plants as responses to a specific set of environmental conditions or questions, if you will. Although so-called improved varieties are superior under today's circumstances, they may not be if there are changes in climate, soil, pest problems, availability of good land, or availability of inputs, such as fertilizer. Having a rich genetic stockpile is essential. It is really just common sense—just as a hand of ten cards is more likely to hold an ace than a hand of three cards, so are plant "aces" more likely to be found if more varieties are preserved.

The Irish Potato Famine is a tragic example of the consequences of too little diversity. The few varieties of potatoes brought from South America back to Europe by explorers took root quickly—by the mid–nineteenth century, the peasant Irish, mostly tenant farmers, depended on potatoes almost completely for food. It turned out the few varieties grown were susceptible to a disease called the late blight, which struck in 1845 and destroyed half the crop. The following year there was a similar crop failure. The result: one million people died; two million more fled the country, many to the United States (a new resistant strain of late blight hit potato crops worldwide in the 1990s).

An outbreak of southern corn blight caused heavy losses among genetically uniform corn crops in the 1970s. Despite such cautionary tales, the appeal of superproduction seems to overwhelm most other considerations. Take corn: Since the commercial introduction of hybrid corn in the 1920s, yields have boomed from around twenty-five bushels per acre to as much as 118 bushels to the acre in the mid-1980s. (A hybrid plant results from a cross between parents that are genetically dissimilar in some way.)

The hybrid revolution in American corn production has spread to other crops and other countries. The so-called green revolution of the 1960s was a transfer of Western industrial agricultural technology to third world countries. Scientists at research labs bred improved cultivars (cultivated varieties) of wheat and rice for tropical areas. These cultivars had a number of characteristics compatible with industrial agricultural practices. They were bred to be fertilized. If this was done, they provided dramatic increases in yield. They were adaptable to various locations and short stemmed, therefore compatible with machine harvesting. By adopting these crops and the techniques of industrial agriculture, countries such as India and Mexico hoped to increase their food supply and better feed their growing populations. And they were successful in this, as were other countries such as Malaysia and Turkey, where grain production boomed.

Thirty years later, the green revolution is the status quo and appears ripe for reform, at least. Some would advocate a counterrevolution because, among other problems, the green revolution's effect on biodiversity has been profoundly negative. Green revolution varieties have replaced local varieties on about half of the wheat acreage and more than half of all rice acreage.[30] Indonesia alone has lost 1,500 varieties of rice in the past several decades. With so much land devoted to these hybrids, there is little room or incentive to grow local varieties. This fact leaves the lion's share of that country's crop uniquely vulnerable to new pests and diseases, without a stockpile of varieties as alternatives.

There are ongoing attempts to save the world's seeds before it is too late. Scientists collect seeds from crops and the wild relatives of crops from around the world and store them in places such as the USDA's National Seed Storage Laboratory in Fort Collins, Colorado, the "Fort Knox of seeds," where samples are kept in fireproof, bombproof vaults with thick concrete walls. Seeds are kept at 35 degrees Fahrenheit and 35 percent relative humidity.

However, the U.S. government has been criticized in recent years for not supplying the Agricultural Research Service, which is responsible for the National Plant Germplasm System (the network of USDA seed banks), with adequate budgets.[31] Seed banks around the world also face a number of problems, including lack of documentation, lack of adequate long-term storage, and lack of duplication.[32]

All of which points to the need for as many people as possible to get involved in saving seeds. A number of private nonprofit organizations are devoted to preserving agriculture's precious biodiversity. The American Livestock Breeds Conservancy (ALBC), mentioned earlier, works with its 4,000 members to save endangered livestock. It researches breed populations, works with breed associations, and educates its members about genetic diversity. Membership is varied and represents a cross section of rural America, including conventional farmers, "ruburbanites" with ranchettes, and new homesteaders in flight from urban America, attempting to make a living and raise children in a healthy environment. In addition, the ALBC has attracted city dwellers that want to lend support to the cause.

The Seed Savers Exchange in Iowa and Native Seeds/SEARCH in Tucson, Arizona, are two organizations that disseminate seeds not readily available commercially. Gardeners and farmers can purchase heirloom, local, and indigenous varieties of vegetables, grains, fruits, and ornamental plants. The Seed Savers Exchange, established in 1975, has about 11,000 vegetable and horticultural varieties in its collection. Through its directories of member seed savers, the organization facilitates the exchange of local, old-fashioned, or rare varieties of vegetables, fruits, herbs, and even flowers.

These organizations are important because, unlike the USDA seed banks, they focus their efforts on farmers and gardeners, adding quite a bit of diversity to the efforts of the people who are working to preserve our heritage of plants and animals. This is important because storing seeds in a seed bank is not enough to preserve biodiversity. Seeds must be planted and crops must be grown, and livestock must be bred and raised in local conditions in order to truly maintain genetic diversity. It is not enough to have a museum of specimen seeds frozen in time. Farmers and ranchers must continue to do what they have done for thousands of years: Select and improve varieties of crops and breeds of livestock that are best adapted to their environment and best fit their needs.

TERMINATOR TECH

The traditional role of farmers and ranchers in preserving the world's agricultural diversity is threatened by certain practices of industrial agriculture and, in particular, the increasing role of biotechnology. Biotechnology carries the trend toward increasing uniformity in crops and livestock one step further—the most infamous example is Dolly, the cloned sheep, genetically identical to her progenitor. Dolly shows what can happen when biotechnology is used to promote uniformity rather than diversity.

One type of biotechnology is genetic engineering, also called gene splicing or recombinant DNA technology. In genetic engineering, genetic material is manipulated or rearranged in order to alter hereditary traits. In short, genes are taken from one organism and inserted into the DNA of another.

Some see this as the logical extension of selective breeding, the ancient process of humans breeding crops for the characteristics they want. It is different in a profound way, however, because genetic engineering is not constrained by the natural boundaries of genus, species, phylum, or kingdom. One could insert a gene from a fish into a tomato, for example.

Although one might be opposed to this practice on a number of grounds—health, environmental, even moral—at first glance, the technology might seem promising for biodiversity. New varieties of crops with desirable characteristics, such as drought resistance, could be created in half the time that it would take with conventional selective breeding.

So far, however, much biotechnology research has been focused on developing herbicide-resistant crop strains. Most products on the market now have been engineered either to tolerate widely used herbicides such as Monsanto's Roundup or to produce their own insecticide, courtesy of a bacterial gene.[33]

One popular new biotech crop is *Bt* corn. This corn produces its own insecticide, *Bacillus thuringiensis,* which kills corn earworms and European corn borers, but it looks like *Bt* corn may have a negative effect on biodiversity. Recently, Cornell University researchers found that pollen from *Bt* corn dusted on milkweed killed or stunted the growth of monarch butterflies. Milkweeds are the exclusive food of the monarch larvae. There is disagreement as to how far the wind might carry corn pollen (anywhere from three to sixty meters) out of a cornfield and therefore disagreement as to the impact on monarchs.[34] But the case of *Bt* corn and the monarch butterfly reinforces the fear of many that bioengineered crops could have unpredictable ecological consequences.

Traditional crop breeding programs have relied on the time element to determine if a cultivar will be useful under various environmental conditions such as drought, high rainfall, and pest challenges. However, often biotechnology companies are in a race for profits and often do not take time for adequate testing.[35]

I think extreme caution should be used when considering the use of bioengineered crops. The effects are already being felt not only on the microlevel, in the field, but on the macro level, potentially affecting all of agriculture—indeed, all of life. Serious social and economic questions are raised by the "biotech revolution." With genetic engineering has come an unprecedented expansion of the concept of ownership: life-forms may now be patented at the level of the gene. These new patenting and intellectual property regulations will permit corporations to continue to freely appropriate unpatented seeds from around the world, to modify a single gene of these seeds, and then patent and acquire exclusive rights over them. These new patenting laws are clearly designed to transfer the ownership and control of the world's seed diversity—much of which has been maintained by traditional farmers in the third world—into the hands of first world corporations. Moreover, seed/biotech corporations have been buying out or taking control of seed banks and smaller seed companies, some say, to reduce the availability of unpatented and nonhybrid seed varieties. It is in the financial interests of these corporations that farmers purchase patented seeds from them year after year.[36]

Serious social and economic questions are raised by the "biotech revolution."

Two strategies are now being used to prevent farmers from saving and replanting their seeds from the previous year's crop, an essential practice in maintaining local biodiversity. It is now possible for scientists to deliberately engineer any crop variety to be sterile or nonreproducible. This "terminator" technology, as critics call it, has been patented in the United States. This takes the responsibility for and opportunity for biodiversity out of the hands of the many, namely farmers and small companies around the world, and puts it in the hands of the few, the so-called "life science" corporations.

To add insult to injury, all patented seeds will now be sterile in a legal sense, as the new patenting and plant breeding regulations give patent holders rights that enable them to prohibit farmers from freely saving and replanting their seeds. To help enforce this, new DNA "fingerprinting" techniques can be used to identify the genetic structure, and therefore the ownership, of crops growing in any farmer's fields. (This begs the question: To what degree does the farmer, who grows such crops, actually own them?)

For the first time in history, farmers are losing both the ability and the right to save and replant their seeds. Yet it is farmers saving, replanting, and crossbreeding seeds that has created the enormous diversity of domesticated crops, and crop varieties we have inherited to this day.[37]Again, which will more likely remain standing if a brick is knocked out—a wall built of just a few large bricks or a wall built of a

**Checklist for Farmers:
How to Encourage Biodiversity**

1. Diversify farm enterprises.
2. Incorporate livestock and pastures into your farm.
3. Manage or seed pastures in order to get a variety of forages, including legumes and native species.
4. Rotate crops and rotate row crops with hay crops.
5. Leave strips of vegetation at field edges.
6. Cut use of pesticides and fertilizers.
7. Plant cover crops.
8. Try intercropping and stripcropping.
9. Grow crop cultivars with diverse genetics.
10. Raise heritage breeds of livestock and poultry.
11. Try conservation tillage.
12. Take marginal land out of production and leave for wildlife.
13. Plant wildlife food plots.
14. Fence riparian areas.
15. Provide corridors for wildlife.
16. Plant trees and native plants.
17. Mow hay meadows after birds have finished nesting.

large number of smaller bricks? "Don't put all your eggs in one basket" is a bit of wisdom from the farm and an old saying that perhaps says it best.

The planet's biodiversity is under assault at the worst possible moment, considering the predicted global warming, the ever-growing population, the losses of prime farmland to urban sprawl, and the shrinking of energy supplies. But with more citizens becoming aware of the problem, the situation is not hopeless. For inspiration, I like to think of the Soviet agricultural scientists who risked their lives during World War II to protect a cache of South American potatoes. In a Leningrad basement, they fended off rats and freezing temperatures and endured hunger in order to save not the crown jewels but a lowly pile of potatoes, which they viewed as a national treasure, an essential for the revitalization of Russian potato varieties, and therefore an essential for national security.[38]

We need a little of that spirit today, at a time when we need more biodiversity than ever if the world is to feed itself.

Chapter 7

The Sorcerer's Apprentice: Step 6—Manage Pests with Minimal Environmental Impact

> For the first time in the history of the world, every human being is now subjected to contact with dangerous chemicals, from the moment of conception until death.
>
> Rachel Carson, *Silent Spring*[1]

On summer evenings, as the sun inched toward the horizon and the air began to cool and the constant Great Plains wind died, we would wait expectantly for the buzz of the crop duster. The other kids in the community and I liked to watch these little planes appear over the fields of cotton, trailing long white chemical tails. The pilots, often veterans of World War II or Korea, were daredevils who seemed to have no fear as they sped across the fields, turning on a dime. With military precision, they would fly down those cotton rows, so low they might catch a few stalks, then pull up to miss the fence but still passing low enough to zoom under the rural electric power lines. The dust or spray would float down on the green cotton plants like fog. If there was any breeze, the chemical would fall on the garden and on us. Before long, the boll weevils and bollworms would be dead, one more hurdle overcome on the road to making a good harvest.

To earn a little bit of extra cash, my father sometimes worked as a marker or spotter for a crop duster named Gunnar Schultz, who was a World War II veteran and a skillful flyer who now lived in our community. A marker's job was to stand, usually at the end of a crop row, so the pilot could center on him; then he would move sixty feet over to mark the new center, and so on, until the job was done. Markers such as my father were invariably doused with dust or spray.

We called in the crop duster only when we didn't have time to do it ourselves from a tractor. Cotton seemed to demand regular dousing with insecticide; we sprayed often with some very toxic chemicals. In the fall, we sprayed arsenic-based defoliants in preparation for harvest. We seldom took safety measures. Because of the sediment-laden well water we used to mix the spray, the sprayer nozzles clogged and we had to clean them often. It was more easily done with our bare hands than while wearing gloves, so that's what we did. I remember my father's fingernails being eaten away and then falling off from contact with the arsenic spray.

In those days, we called the array of 'cides—insecticides, fungicides, herbicides, miticides—by a simple name: "poison." Despite our understanding of the nature of the beast, we still did not consider the poison a risk to us. No one told us the sprays were dangerous. After all, they were made to kill insects, weeds, or fungi, not people. And they were diluted with water. How dangerous could they be?

This was before Rachel Carson warned us of their dangers in *Silent Spring,* before the Environmental Protection Agency was even thought of, in an age when everyone still believed in better living through chemistry. We accepted spraying as part of the business of growing cotton. I remember the local conventional wisdom that said if you started spraying, you couldn't quit. (In other words, total annihilation of the pest was the Holy Grail.)

Furthermore, as farmers in southwestern Oklahoma, we were grateful that these sprays helped us make a living in a place where making a living was not easy. Farmers here spent their lives trying to overcome weevils, hail, drought, low prices, and swings in government subsidy programs, all while living in tornado alley. Snyder, a nearby town, was dubbed "Cyclone City" because it had been nearly decimated twice.

This kind of living makes people tough. We could be knocked down, but we wouldn't stay down. Although this is laudable, frontier culture has its downside. When trouble, in the form of the boll weevil, came around, we fought back hard in the only way we knew how: with strong chemicals. And we grew callous to the effects. One day, my grandfather and I walked the two miles down to Otter Creek to fish. We arrived at a good spot on the bank only to find the surface of the creek dotted with dead fish. I don't know for certain what caused the fish kill, but we could smell the insecticide in the air from the adjoining cotton field.

At the time, I viewed the incident as a curiosity. No one thought much about pesticides entering the water supply, and even when there was a fish kill, no one was very concerned. We fished in the creek, fish kill or no fish kill. Now, when I think about the way the Otter Creek bottom reeked from the smell of chemicals and the look of those dead fish, it makes me sick. And, in the end, all the spraying we did to make a crop each year didn't save our farm or the farms around us.

And it didn't provide my father with a contented old age. In 1967, he complained of feeling weak. After several days of this, he went to the doctor who put him in the hospital. There he was diagnosed with acute leukemia; he died three days later. He was forty-two years old.

FROM CONCEPTION UNTIL DEATH

The boll weevil crossed the Rio Grande River and invaded the cotton fields of Texas in 1894.[2] It quickly spread, prompting some Texas farmers and businesspeople to place a bounty on the head of the insect, paying from ten to twenty-five cents per hundred.[3] The weevil spread across the South with devastating effect. Typical was the plight of a planter in Louisiana who had been harvesting 500 bales from his fields, who the next year harvested sixty-three, and the next only twenty-five bales after the weevil arrived.[4]

The boll weevil, nemesis of my childhood, is a good example of how devastating insects can be to crops and rangelands. Bacteria, fungi, viruses, and nematodes, spreaders of the some 50,000 known plant diseases,[5] add to a farmer's woes.

Then there are weeds. Weeds compete with crops for sunlight, space, nutrients, and moisture. Some weeds, notably lamb's quarters and ragweed, are incredibly thirsty and can absorb twice as much water as some crops.[6] Competing with more desirable plants, weeds reduce crop yields in the United States by ten billion dollars annually.

It's no wonder that farmers embraced chemical pesticides in such a big way after World War II. Using them was so easy. Before then, farmers relied on cultural practices such as crop rotation, tillage, companion plantings, and mechanical removal to fight insects, diseases, and weeds. Only a few pesticides, such as nicotine, arsenic, and pyrethrum, were available.

From 1964 to 1982, the amount of pesticides used in United States agriculture more than doubled.

Today, many more pesticides are now available—278 used directly on raw agricultural crops[7]—and their rate of usage has skyrocketed. At first, pesticides were used occasionally to prevent a catastrophic loss of a crop; as time went by, they were used routinely to prevent pest populations from building, even if damage was not imminent.[8] From 1964 to 1982, the amount of pesticides used in United States agriculture more than doubled.[9]

For a decade or so after 1982, the use of pesticides, measured in pounds of active ingredients applied, seemed to have stabilized, but this was not due to farmers finding alternatives to pesticides. Analysts say farmers switched to more potent and persistent products that work at a lower dose, so that fewer pounds used does not necessarily correspond to a lesser threat to the environment.[10] However, in 1995, the numbers were up again, to 966 million pounds according to the EPA.[11] If divided equally among the roughly two million farm operators, that would be about 480 pounds of poison per farmer.

On a summer evening drive through miles of cornfields, one's horizons are filled with corn: the smell of it, the rustle of its leaves, and the bobbing of tassels in the breeze. If there is a quintessential American movie farm scene, this would be it. Unless one happens upon a spray rig, what the casual observer usually doesn't experience directly (and what the movies don't show) are the pesticides that have been applied to those fields of corn. In the major farm states in 1995, pesticides were applied to nearly all fields of corn—as well as of soybeans, cotton, potatoes, and spring and durum wheat.[12]

Pesticides are defined as toxic chemicals deliberately used to control plant and animal pests. Using pesticides to kill pest organisms has been gospel since World War II—an approach called the "therapeutic" model of pest management; in other words, treating the problem after it occurs.[13] The creativity and resources of the American chemical industry have been unleashed in support of this paradigm.

Agricultural pesticides are often classified into three groups: insecticides, herbicides, and fungicides. Which pesticides are used depends on the crop. Insecticides and fungicides are sprayed more heavily on fruit and vegetable crops; insecticides are sprayed heavily on cotton, as they were when I was a kid. But it is herbicides, chemicals that kill weeds, that are by far the most used in agriculture; they constituted 63 percent of the total weight of all active ingredients in 1995.[14] Herbicide use has exploded over the years—in 1964, they accounted for only about 25 percent of pesticide quantity.[15]

All these millions of pounds do not come cheap. Americans spent 10.4 billion dollars on pesticides in 1995, three-quarters of that for agricultural use (about 7.9 billion dollars).[16] Some would consider that a good deal. Proponents argue that pesticides are effective, and they are less expensive than labor-intensive pest-control practices.

However, both of those assertions are becoming less true. Take the first: effectiveness. More than 500 insect pests, 270 weed species, and 150 plant diseases are now resistant to one or more pesticides.[17] The result is that we are losing about one-third of our crops to pests in the United States,[18] virtually the same percentage as before pesticides began to be used heavily post–WWII. Although pesticides were spectacularly effective when first used, before long, target insects developed resistance to pesticides. At the same time, pesticides killed off nontarget pests that were often beneficial predators, the natural enemies of the pest. The result has been a pest rebound.

As for cost, it too is rising. In agriculture, it rose almost 35 percent from 1983 to 1993.[19] Expenditures for pesticides now account for a significant percentage of farm budgets, about 17 percent of variable cash expenses for corn, to 25 percent for cotton in the southeast and delta regions.[20]

In addition to these direct costs, there are hidden costs for using pesticides, including the many risks to both the health of people and the environment. And although agricultural economists don't generally include these risks in their calculations, common sense says they should be included. Common sense and the facts tell me that the way we have been using chemical pesticides is not healthy and, in the long run, not sustainable.

THE KILLING FIELDS

The way pesticide use has evolved is a little like the scene in Walt Disney's *Fantasia,* the animated classic. When the sorcerer leaves his workshop, his apprentice, the naive Mickey Mouse, is tempted to try out some of the sorcerer's powerful spells. He dons the sorcerer's pointed hat and casts a spell on the broom. The broom comes alive and begins doing Mickey's work for him, hauling water from a well to fill up a tub. Soon the loafing Mickey is lost in his daydream of flying among the stars and then falls blissfully asleep. Meanwhile, the broom keeps on working, and soon the tub runs over. When Mickey finally wakes, the water level in the room is rising, and he commands the broom to stop. But he soon finds out he is not powerful enough to stop what he has started, so, in desperation, he smashes the broom into pieces. Unfortunately, each broken piece comes alive, and the unrelenting army of brooms continues to haul water. The workshop is being flooded.

Mickey has created a monster that he cannot control. What started out as a good thing becomes a scene of chaos and destruction. Luckily, the sorcerer comes back, plucks the hat from Mickey's head, and restores order before the workshop is ruined.

We are, I fear, like Mickey—in over our heads in this workshop known as Planet Earth. The truth is, what seems so helpful and easy, such as spraying to control a pest or two, often has unforeseen consequences. Pesticides, like the magic brooms, do what we want: They kill pests. But they turn out to be not so easy to turn off or control. They often have undesirable effects on organisms besides the ones they target—soil organisms, nontarget insects and plants, birds, fish, and animals. And, of course, humans: Pesticides can make people sick, or even kill them.

From a production standpoint, when resistance to pesticides builds, it renders them ineffective. We must then produce new pesticides, until these don't work or their side effects are discovered. Then we make new pesticides, ad infinitum, tying the farmer ever more tightly to chemical companies and their expensive products and altering the environment ever more radically. Unfortunately, there is no sorcerer who can set things right with a wave of his wand. We are the ones who must figure out how to restore a more natural order.

The truth is, what seems so helpful and easy, such as spraying to control a pest or two, often has unforeseen consequences.

What are some side effects of pesticides? Below ground zero, in the soil, pesticides can kill the essential microorganisms that decompose organic matter, thus interfering with natural fertility cycles. Insecticides can kill earthworms, soil arthropods such as mites, and beneficial ground beetles, all of which have crucial roles to play in maintaining healthy soil. A study of the long-term effects of one herbicide, atrazine, in an orchard found that yearly applications of the chemical caused significant negative soil changes.[21] Even the relatively benign glyphosphate, the chemical in the widely used herbicide Roundup, can keep legumes from fixing nitrogen and can harm mycorrhizal fungi.[22] These fungi make the essential plant nutrient phosphorus available to plant roots.

Aboveground, broad spectrum insecticides do what is advertised, which is to kill a broad array of pests. They also kill insects that are not pests. Many insects, for instance, are beneficial—acting as pollinators or predators of harmful insects. In many instances, pesticides have killed "good" insects, with disastrous effects. One illustration: California olive growers wanted to protect their groves from the destructive olive Parlatoria scale. Normally, wasps controlled 70 to 90 percent of the scale. The growers sprayed their groves with DDT in an attempt to kill the remaining scale; unfortunately, the wasps were killed instead, causing the amount of scale to explode and seriously damage the olive crop.[23]

Pesticides can kill a variety of nontarget species such as fish, birds, frogs, salamanders, and, as above, beneficial insects, including bees, which are essential to the pollination of many crops. Stories of the unintended effects of pesticides abound. In the early 1990s, a herbicide used to control weeds in rice fields in California found its way into the Sacramento River, where it killed large numbers of fish and other aquatic organisms.

Probably the most infamous pesticide is DDT. DDT and other synthetic organic insecticides made of chlorinated hydrocarbons launched the postwar pesticide era. They were originally developed from chemicals tested for potential use as nerve gas weapons during World War II. Developed in Switzerland in 1939, DDT was considered a boon, successfully used in the third world to control malaria, yellow fever, and typhus.

By the early 1950s, DDT had been adopted enthusiastically by American farmers because it was cheap, easy to dissolve and spread, toxic to insects, and, most important, fast acting—almost instantly and highly (90 percent) effective against big outbreaks of target insects.[24]

The problem with DDT was that it didn't go away. It remained in the environment to be ingested. The higher up the food chain, the more concentrated in tissue it became, with terrible effects on predator species at the top. Forty species of American birds, including the bald eagle, were hit hard by DDT and other chlorinated hydrocarbons. The chemical interfered with calcium metabolism in these birds, and they began producing soft-shelled or shell-less eggs and malformed young; their numbers declined rapidly. Our national symbol had to be put on the EPA's endangered species list.

Although banning the use of DDT in 1969 (and other chemicals like it in 1972) helped restore the bald eagle and other birds to health, almost thirty years later it is still with us, like an unwanted tenant we cannot evict. Twenty years after it was banned here, the U.S. Geological Survey has found that DDT and chemicals formed from its breakdown, DDE and DDD, are still present in the Yakima River Basin. Levels of these chemicals are still elevated in farm soils, stream water, suspended and streambed sediment, and fish in the basin. This poses a continuing threat to fish and birds.[25]

Our national symbol had to be put on the EPA's endangered species list.

It also poses a threat to humans, because DDT is carcinogenic. In fact, 107 of the active ingredients in pesticides have been found to cause cancer in animals or humans. Despite this, eighty-three of the 107 are still in use today. According to the EPA in 1993, it takes ten years to ban a pesticide in the United States using present procedures.[26] DDT is still used in foreign countries and finds its way back here in residues on imported food.[27] Other toxic pesticides are also exported from the United States, some at an ever-increasing rate. This includes pesticides that have been banned here, were never registered here (never evaluated by the EPA), and others designated "extremely hazardous" by the World Health Organization.[28]

Trading Health for Production

Just how dangerous agricultural chemicals are and how much we should worry about them is a hot topic of debate these days; indeed, it has been for decades. The debate has gone back and forth. Lately, a spate of books has appeared trying to debunk "hysterical" claims about pesticides, while assuring readers that pesticides—particularly pesticide residues in food—are the least of our worries.

These authors correctly point out that smoking tobacco is the leading cause of cancer in the United States. They also assert that eating lots of fruits and vegetables (even containing small amounts of pesticide residue) can protect a person from cancer. Although it's not clear what the risks of ingesting small amounts of pesticides over a long period of time might be, usually these critics speak only of the general population and don't address the much greater risks to children, who have developing organs and much lower body masses than adults. These commentators also ignore the risks to those who come into regular contact with these pesticides, such as farmers.

In their willingness to ignore health risks to farmers, they are not alone. Plenty of farmers today downplay the risks to themselves. Instead, they complain bitterly about the regulation and restriction of the pesticides they rely on.

The truth is, industrial agriculture does depend on hazardous chemicals. Without them, many farmers believe they could not make an acceptable profit. The attitude was summed up by Victor Davis Hanson in his book *Fields Without Dreams*. He describes the California raisin growers in the 1970s when credit was readily available and prices were up:

> A new generation of very potent pesticides and herbicides was added to the wealthy raisin grower's already lethal arsenal. Every living thing—worms, insects, fungi, viruses, and bacteria, in the soil, on the ground, on the vine, in the air—was to be targeted. Forget whether expensive chemical vineyards are sustainable in the long haul. Ignore your own family's drinking water supply

> pooled not far below in a subterranean lake beneath your tractor. Turn a blind eye to your son on the daily spray rig.[29]

Davis speculates whether "the ugly looking growth" on the neck of a neighbor's son was caused by five years of daily fungicide dusting with the "powder on his clothes, in his truck, powder on his tools for weeks afterwards. . . . In the short term all that mattered was that these killing fields produced a lot of raisins."[30]

Clearly the concept of the farm as a part of nature is not part of the above scenario. Just how dangerous are such chemicals to farmers, farm workers, and people in rural communities? Recent estimates range from 10,000 to 20,000 deaths from pesticide poisoning each year worldwide, and from three to twenty-five million acute severe cases of pesticide poisoning occur, many in developing countries.[31]

Whatever the exact numbers, the World Resources Institute asserts that "exposure to pesticides with known human toxicity continues to be widespread and heavy in many countries around the world. . . . Direct observations and biological measurements bear this out."[32] Recently other scientists have concluded, "pesticide risks today are at least as serious as they were in the early 1970s. In addition, overall risk and some specific types of risk appear to be currently increasing."[33]

Farmers and farm workers have myriad opportunities to come into contact with pesticides: while mixing, loading the sprayer, applying, and then entering the field after application. Accidental spills and improper disposal can also expose farmers.

Since the 1970s, other classes of insecticides, organophosphates (such as parathion) and carbamates (such as Sevin)—have largely replaced the chlorinated hydrocarbons. The good news is that they are short lived and not prone to concentration in the food chain. The bad news is that some are more acutely toxic and kill millions of fish and birds each year.

Organophosphates and carbamates act to inhibit cholinesterase, a nerve enzyme. Sometimes the effects are subtle. In one study, crop dusters who applied organophosphates were found to suffer from loss of memory and lack of concentration. Flu-like symptoms can signal mild poisoning. A 1978 University of Nebraska study found that 30 percent of farmers and commercial pesticide applicators had a reduction in cholinesterase levels and almost one-quarter had symptoms of mild poisoning. None had complained to a doctor.[34]

Recent estimates range from 10,000 to 20,000 deaths from pesticide poisoning each year worldwide, and from three to twenty-five million acute severe cases of pesticide poisoning occur, many in developing countries.

Pesticides may also cause health problems over the long term, perhaps from repeated lower exposures. Besides the risk from any given pesticide, we are at risk from exposure to combinations of toxic chemicals, whose effect may be amplified in combination. It is known that pesticides can affect the endocrine, nervous, immune, and reproductive systems.[35]

Many pesticides in use have been shown to be carcinogenic in animals. But there are other ways to assess the risk to humans. Epidemiologists use statistical methods to try to establish links between diseases, such as cancers, and their possible causes. This is often the only way when there is a long lag time between cause and effect or, as is the case with pesticide exposure and cancer, other influencing factors such as individual physiological differences, or unknowns such as the effect of cumulative exposure. As analysts at the Center for Rural Affairs in Nebraska have pointed out, "If enough studies show people who use chemicals die of particular causes more often than other people, eventually we can conclude, as primitive people concluded that water makes plants grow, that the chemicals are somehow related to the deaths."[36]

Looking at such studies is probably the closest I will ever get to knowing whether my father's leukemia was caused by his exposure to pesticides. Studies of farmers both here and abroad found that we have increased risk of cancers of the primary central nervous system, lung, and lymph nodes. A 1986 study by the National Cancer Institute found that Kansas farm workers exposed to herbicides more than twenty days per year had a six-time-higher risk of developing non-Hodgkins lymphoma than nonfarm workers. Specifically, agricultural use of herbicides such as 2,4-D has been associated with two- to eightfold increases in this cancer in studies conducted in Sweden, Nebraska, Canada, and elsewhere, according to a study published in 1992 in the journal *Cancer Research*.[37]

Most telling for me, in thinking about what may have caused my father's death, are studies in Nebraska and Iowa that found that people living in counties with high herbicide and insecticide use were significantly more likely to die of leukemia.

The reasons for these elevated risks of cancer may lie in the way that pesticides affect the immune system. Experimental animal and wildlife studies as well as human studies point to pesticides as the producers of some "significant changes in immune system structure and function." In medical parlance, they are immunosuppressive. These changes are "accompanied by increased risks of infectious diseases and cancers . . . even in otherwise healthy populations."[38] Cancers of the immune system such as leukemias, lymphomas, and myelomas occur more commonly in immunosuppressed people, such as AIDS patients, and those exposed to pesticides, such as farmers.[39]

Having a depressed immune system is particularly perilous for people in developing countries who suffer from lack of sanitation, contaminated water, and crowded, poor housing. However, it's not just in third world countries where people are at risk down on the farm. According to the U.S. Department of Labor, American farm workers have the highest rate of chemical-related illness of any occupational group.

How Nonfarmers
Can Support Healthy Pest Management

Farmers often complain that they are unfairly singled out for excessive application of pesticides, and they have a point: Consumers apply pesticides in and around their homes four billion times a year![40] Consumers also have a direct impact on how much pesticide is applied, postharvest—an estimated 10 to 20 percent of insecticides used on fruits and vegetables in the United States are applied solely so consumers may have perfect-looking produce.

1. Buy organically grown, transitionally grown, or sustainably grown produce and meat in your farmer's market, supermarket, or directly from a farm. If your grocery doesn't carry such produce, request it.
2. Don't insist on cosmetically perfect produce. Often the most perfect looking is the most sprayed. Develop an understanding of how food is produced.
3. Don't automatically spray when you see a bug on a plant in your yard or garden. You may be seeing a beneficial insect. Killing that insect may throw the system out of balance and contribute to a buildup of bad bugs.
4. Look into alternatives for common household pest problems. Many are cheap and effective: for example, using boric acid for roach control.
5. Recognize that your lawn does not have to be a monoculture of one type of grass. Tolerate weeds—they are green, can be mowed, and really don't hurt anyone.
6. Plant trees and shrubs around your home that are native or well adapted to your climate and soil. Doing so cuts down on disease and pest problems.

THE CHEMICAL TREADMILL

Despite the growing body of evidence on the health perils of pesticides, some still argue that farmers cannot do without such chemicals. In fact, critics of pesticide use have often been dismissed as reckless advocates of turning back the clock to the days before pesticides, and therefore dooming the world to mass starvation. I was once like others in agriculture who shared the mind-set of Nixon-era Secretary of Agriculture Earl Butz, who warned that without chemicals, tens of millions of people would go hungry. (I suppose the unspoken reasoning was that thousands of farmers around the world may have to die each year to ensure the planet's food supply.)

I no longer accept those dire predictions. In fact, some estimate that if nonchemical controls replaced insecticides completely, only 5 percent of food production would be lost in the United States. Reduction of pesticide use in some countries, such as Indonesia, have resulted in better, not worse, yields.

The time has come for a transition to a new kind of pest control. I do not advocate abandoning chemicals overnight: the socio-economic-political system could not stand going "cold turkey" like that. But I think reductions have been proved possible and research into alternatives needs to be of the utmost priority. Furthermore, we need to vigorously encourage farmers to use alternatives that work. Certainly for humanitarian and ecological reasons, and increasingly, for economic ones, too.

In short, pesticides are not working as well as they used to. One big problem, as mentioned earlier, is resistance. When chemicals are applied, there will likely be individuals, be they bug or weed, that are resistant to the chemical's effect and live to reproduce. Eventually, whole populations will develop a resistance to the pesticide.

The pace can be dramatically swift—the number of weed species resistant to herbicides has jumped from forty-eight to 270 just in the past ten years. Resistance even affects animal agriculture—the use of ear tags to kill fleas and ticks on cattle can lead to resistant species. As pest populations build resistance, the effectiveness goes down, causing some farmers to up the dosage or spray more frequently, which simply accelerates the process (and increases costs). Chemical manufacturers then attempt to produce a different chemical to meet the need. This has been called the "chemical treadmill." Once a farmer gets on, it's hard to get back off.

A treadmill is an apt metaphor, because, viewed on a macro level, using pesticides can be like walking but never getting anywhere. You feel as if progress is being made, but it is an illusion. Wipe out one pest, and a new one moves in to fill the niche occupied by the old pest. Wipe out target pests, their natural enemies go with them, and you make the problem worse. Spray an herbicide to kill weeds, and fungus diseases and insect problems can worsen. Spray an insecticide and it drifts over and kills the colonies of honeybees on your neighbor's land, so his fruit trees don't get pollinated.

Pesticide fallout reaches beyond farm communities, of course. Pesticides find their way into water supplies and can accumulate in human fat, even among people far from exposure to farm fields. The effects of pesticide residues in food on human health are still being debated: questions remain about what are safe levels, and what are the long-term effects of ingesting small amounts of pesticides over a lifetime. In fact, the EPA is currently reassessing 470 pesticides for health risks to infants and children, for effects on the human endocrine system, and for possible cumulative effects.

As if pesticides used to produce food weren't enough, some pesticide residue in our food comes from postharvest applications. Government standards on the cosmetic appearance of foods compel some extra pesticide application, as do consumer demands for perfect-looking food. Interestingly, pest damage one can see is not tolerated even if it poses no health threat, yet unseen residues, which might be harmful, are better accepted.

All these concerns aside, here is the bottom line: although pesticide use has risen since the 1970s, crop losses to pests have not declined. Worldwide, the pattern is similar to that in the United States: Crop losses due to arthropods, diseases, and weeds have actually increased from 34.9 percent in 1965 to 42.1 percent in 1988 to 1990.[41]

This while chemicals eat up sizable chunks of farmers' money. Add to these direct costs the indirect costs of human poisonings, fish kills, honeybee losses, and bird and mammal poisonings, which have been estimated at one to three billion dollars annually. So it's not surprising that, in the Philippines, the International Rice Research Institute found that when health costs are counted as a production cost, the use of pesticides in that country cuts, rather than increases, rice productivity.

The costs to the public of agriculture's current heavy reliance on pesticides are extensive. Besides health care costs for farmers, farm workers, and others in the general public who become sick from pesticides and the cost of cleaning up pesticide pollution, there is the cost of government regulation. Who bears these costs? It is not, by and large, the chemical companies that produce the pesticides. It is the ordinary taxpayer who bears the burden to a large extent, by paying for such programs as Medicare and Medicaid, the Environmental Protection Agency, and other government agencies charged with protecting our health and the health of the environment.

Of course, many producers, especially those with small or medium-sized farms, operate on the edge of profitability. Asking them to bear all of the costs would be unfair and, in many cases, financially disastrous. Ultimately, consumers and others in the food processing and distribution chain must be made aware of the total cost of producing food and be willing to pay for it—both directly and indirectly—through research on ways to farm without chemicals.

The good news is that farmers and scientists are today researching many exciting new approaches to pest control that are healthy, are effective, and will, it is hoped, stop the treadmill. Some are quite simple yet effective: For example, Turkish researchers have found that the vapors from the essential oils of cumin, anise, and oregano can kill pests in infested greenhouses, while apparently posing no threat to workers. Call it aromatherapy with a twist.

Outsmarting the Pests:
Rotating—Attracting—Trapping

Often creating the kind of balanced agroecosystem that enhances natural controls on pests requires adding plants or animals to the mix on your farm.

Crop rotation is one tried-and-true way to do this. Crop rotation is defined as the systematic changing of crops grown on the same land to help prevent soil exhaustion. Besides helping build healthy soil, rotating crops can also help to control pests. Although a continuous planting of the same crop in the same field can cause an increase in insects, nematodes, and diseases, a three-to-four-year rotation of a nonhost crop can reduce these problems. Insects, nematodes, and diseases that complete their life cycles in one to two years are the most susceptible to rotation.[42]

(continued)

(continued)

Why? Insects often overwinter as eggs or pupae in the soil, in cocoons on the host plant, or as adults on plants or weeds around the edges of a field. Often insects overwinter within a few inches to a few yards of where they last fed. The Colorado potato beetle is a good example: If its host plant is not there when it emerges in the spring, it will die. So rotating the host crop with another can stymie such pests. Some crops control weeds by smothering them with rapid growth. An example is the velvet bean, which used to be widely grown in rotation with cotton.

Sometimes cultural weed control melds into biological insect control. Cover crops, which can shade out weeds, are a case in point. Cover crops are often crops growing close to the ground, usually grown to protect the soil from erosion. They are planted between trees and vines in orchards and vineyards or when fields are not being used for main crops. They can attract beneficial insects to orchards and fields. At a research station in West Virginia, researchers have planted a diversified orchard of apple and peach trees and a number of ground covers. Some of these plants act in a fairly straightforward way—rape is toxic to nematodes (microscopic, soil-dwelling pests). Others act in a more indirect manner. Buckwheat flowers provide nectar and pollen for beneficial insects, one of which is a parasite that feeds on the codling moth, a serious apple pest. According to researchers, the quality of the fruit harvested was comparable to fruit grown with conventional pesticide controls.[43]

When such cover crops attract pests, they are referred to as trap crops. How they work is often subtle. For example, it has been discovered that the oil radish and white mustard plants can control a pest of the sugar beet by 80 to 90 percent. They are planted in the field and then plowed under before sugar beet planting. The radish and mustard are so similar to the sugar beet that the pest feeds on their roots, but they are just enough different to keep most of the bugs from reproducing before they die.

Sometimes trap crops are planted adjacent to, around the borders of, or even interspersed with the main crop. One recent SARE project demonstrated that planting black-eyed peas in pecan orchards can help growers manage stinkbug, which can cause major losses in orchards. The bugs seem to prefer the black-eyed peas to pecan kernels. The results were good—for each dollar spent on establishing and maintaining the trap crop, about nine dollars of kernel damage was prevented.

Other sustainable approaches to pest control are less complicated, involving relatively simple cultural practices such as timing a crop to give it a competitive advantage over weeds, or to avoid peak insect injury. Another practice is increasing the density of the crop to shade out weeds. These "smother crops" traditionally include alfalfa, foxtail millet, buckwheat, rye, sorghum, sudan grass, sweet clover, sunflower, barley, corn, and cowpeas.

IT'S NATURE'S WAY

A survey of recent news items tells the tale: Working with nature to control pests is an idea whose time has come. In Hawaii, sugarcane farmers are releasing parasitic insects to control weeds and insects that attack sugarcane. Researchers in the United States find that extracts from the neem, a common tree of India and Africa, can safely kill more than 200 species of insects, and fungus diseases to boot. In Tennessee, weevils are reducing dense stands of musk thistles, a weed that has taken over productive crop- and pastureland. In California and Arizona, bacteria are controlling root rot in tomatoes. In Washington apple orchards, farmers are "messing up" the sex life of the codling moth. How? By releasing sterile males and by hanging plastic strips that smell like females in the trees, thus reducing the numbers of apple-chewing larvae.

Back at the Kerr Ranch, we have had geese eating crabgrass in our organic strawberry patch, and sheep eating the weeds that cattle won't eat. In our laboratory, we have found that extracts of black locust and black walnut leaves can effectively inhibit some seed germination and growth, acting as a natural preemergent herbicide.

These are just a few examples of alternatives to the chemical control of pests, in this effort to cut or eliminate the use of potentially harmful and definitely expensive chemicals. Although it has not been silenced, the buzz of the crop duster is getting fainter.

The first step in the development of such creative approaches to pest control is a change in attitude. Understanding weed and insect species, their characteristics, and how they fit in the agroecosystem are essential steps to this change. Weeds and insects are part of the agroecosystem, and, although farmers and people in general don't think of them this way, they can have a positive role. Weeds can cover bare soil and prevent erosion. They can harbor beneficial insects. They can provide food for birds (ragweed, for example, is an important food source for quail). They can pull minerals from subsoil areas up into crop root zones, which is sometimes very important in providing plant micronutrients and trace elements.

Weeds can also be crops. Crabgrass, for instance, can be an important forage for livestock and, when managed correctly, it is comparable to the highest quality forages. Even the kudzu vine, which some call the curse of the South because of its tendency to cover up whatever it comes in contact with, has its positive side. It's a legume, and thus adds nitrogen from the air to the soil. The Japanese use it to make paper, fabric, and a cooking starch. Some research has shown that compounds from the plant can help reduce the craving for alcohol among alcoholics. Some people actually grow it: one family in North Carolina feeds the high protein (higher than alfalfa) plant to their dairy cows and also makes it into quiche, jelly, and relish.

Although it has not been silenced, the buzz of the crop duster is getting fainter.

As a boy, I gathered careless weed to feed our hogs. Nothing was wasted. My mother canned lamb's quarters to keep us in winter greens. By eating lamb's quarters, we were unwittingly following a very ancient tradition: a species of lamb's quarters *(Chenopodium)* was domesticated and eaten at least 3,500 years ago in eastern North America. It was one of several early domesticated plants that some scholars say established eastern North America as one of the seven primary areas of agricultural development, along with the Fertile Crescent of the Near East and Central Mexico.[44]

Insects, too, have an important place in nature. They are premier pollinators. Without them, there would be no berries, peaches, apples, and other tree fruits, no cotton, tea, coffee, chocolate, clover, or alfalfa. Livestock forage would decline. And without insects, what would become of the birds, fish, and mammals that depend upon them for food?

Not only that, but "beneficial" insects act as predators upon pest insects and therefore provide a kind of natural control of them. Some spiders, praying mantises, and parasitic wasps are examples. Another is the ladybug, or more correctly named, the lady beetle. The whimsical ladybug, beloved of children everywhere, is actually a hungry predator of soft-bodied insects. Its larvae have been called ferocious and insatiable.

The fact that these natural controls exist provides an alternative to the therapeutic approach to pest management. Called the total system approach, it advocates "a shift to understanding and promoting naturally occurring biological agents and other inherent strengths . . . and designing our cropping systems so that these natural forces keep the pests within acceptable bounds."[45]

What are these natural regulators? One example is the natural defenses inherent in the plant. For instance, it has been discovered that when certain caterpillars feed on the leaves of a tomato plant, it induces the production of a substance throughout the plant that interferes with the digestion process and feeding behavior of other insects. Thus, the fruit, the most important part of the plant because it holds the seeds, is protected. Agriculturists can learn from such natural defenses.

Other components of the agroecosystem also have important parts to play as natural regulators of pests: mixtures of plants, the soil, and natural enemies. As others have noted, organisms live in communities, and it is the complex interactions in these communities that provide the most reliable mechanisms for stability.[46]

A healthy, stable farm in biological balance therefore requires a diversity of plants and animal communities. Using this model of a healthy farm, the worse possible scenario for a natural balance and control of pest species is the most common agroecosystem in industrial agriculture: the monoculture. Hundreds—sometimes thousands—of acres of identical plants provide the ideal conditions for gigantic outbreaks of insect pests, which usually leads to spraying of chemical pesticides. In the long run, such spraying can make the situation worse.

That is why sustainable agriculture stresses that for best management, the farmer must look into what caused the emergence of the insect as a pest. If the overpopulation of a pest is the result of the agroecosystem being out of balance, then the farmer must explore how the system is out of balance. If it is a monoculture, plant diversity may not be adequate to provide habitat for beneficial insects and for other natural checks and balances. The crop may be under stress from bad weather or bad soil, making it more susceptible to pests (monoculture also tends to produce bad soil, as discussed earlier). Some contend that stressed plants actually look and smell different to insects than healthy ones do.

Even if plants are healthy, however, they can be afflicted by pests. Under those circumstances, a sustainable pest-control program would, at minimum, develop biological and cultural control of pests (using chemicals as a last resort). This approach minimizes damage to the environment and improves safety for those handling chemicals, while also lowering input costs.

THE TEAM APPROACH

The term *pest management* is used so much in agriculture circles these days that its significance has perhaps been lost. We don't say pest elimination or pest annihilation. We say "management." It implies control, but not complete control. It implies supervision—overseeing, looking over. It implies even a bit of cooperation. In fact, the approach today has something in common with those new team management models so popular in progressive business circles. And, it's true, many of these new pest-management practices in fact forge a partnership between farmer and bug.

How do you manage pests without chemicals? Biological approaches are probably the most publicized pest management alternatives to chemicals. Biological controls include "beneficial insects," including predatory and parasitic insects, fungi, viruses, nematodes, and bacteria. Allelopathic compounds (toxins from plants) and grazing animals can be included in a broad definition of biological control. Sometimes the use of sex pheromones, sterile insects, and resistant varieties are also included under this category. A biological control agent is usually a natural enemy of the pest.

Using these biological agents demands an understanding of their niche in the entire farm system. This kind of research is, to me, the most promising for development of a sustainable agriculture.

Biological approaches are often fascinating and novel. A case in point is the device that fits on a beehive that holds a talc/virus mixture. As bees come out of the hive they are dusted with the stuff, then, as they pollinate blossoms of crimson clover, they distribute the deadly mixture that kills more than 75 percent of earworm larvae found there. Earworms are a pest of corn and cotton.

Sometimes reading descriptions of how these "beneficials" work is like reading science fiction—a world where threatening bugs turn their enemies to mush. This is literally the case with a virus that liquefies armyworms, a cotton pest. Reportedly, "the virus replicates millions of times inside the worm's body, turning the pest's tissues into a dark, slushy mess."[47] The effects of parasitic insects, often tiny wasps, but also including a few species of beetles and flies, are equally grisly—a "helpful wasp," the *Lysiphlebia japonica,* attacks cotton aphids by injecting eggs inside them. The hatched larvae feed on the pest from the inside and kill it.

The respect that beneficial insects are being given these days is reflected in farm magazine advertisements for pesticides which emphasize how hard they are on pests, but how kind they are to beneficials. Impressive research projects, many through SARE, are looking into how to utilize beneficial insects. One study in Florida looked at predatory spiders (arthropods) and their effects on certain pest mites on Florida citrus. The study was, to the layperson, mind-boggling in its painstaking approach. Researchers collected and identified the arthropods, identified their habitat, studied their life cycle, their food, and the effect of different pesticides upon them at different stages of their lives. Because researchers found that many pesticides were highly toxic to these beneficial spiders, they felt their study would "radically change" the way Florida citrus growers approach pest control.

Biological approaches are often fascinating and novel.

Organic farmers, of course, have long relied on biological controls of pests. Another SARE project assessed the impact of beneficial insect populations on organic farms in the South. They found that biological control by naturally occurring parasites and predators, including Trichogramma wasps and ladybugs, was very important in suppressing insect pest populations and helped make organic tomato production economically viable.

One of the more spectacular examples of using insects to do what pesticides used to do was the mass introduction of two types of plant-feeding weevils in several southern states to attack the musk thistle, an introduced plant that had become a troublesome weed on croplands and pasturelands. It has been quite effective, reducing densities by as much as 97 percent at some sites. Researchers estimate herbicide use will be reduced by a million dollars or more per year in some states.

Beneficial insects are usually present in areas where few pesticides have been used. Farms are usually not these places, so farmers often must attract beneficials by providing them a favorable habitat, or, as in the case of the thistle-eating weevils, by introducing them. Too often, say critics, instead of trying to augment natural indigenous populations of natural enemies and understand how they function, predators are imported and used, in effect, as a natural pesticide to cure the problem.

This is also a downside to the new biopesticides—that is, pesticides "made from nature," such as *Bt, Bacillus thuringiensis,* a bacteria used against the destructive larvae of moths and butterflies. Using such naturally based pesticides is not a final answer—pests can become resistant to biopesticides just as they do to chemical pesticides.

Even so, such products are usually much safer for farmers and kinder to the environment. They are a valuable tool and supplement as we are learning how to design more self-sustaining agroecosystems that do not require "therapeutic" interventions.

I just hope we remember our ultimate destination and don't get sidetracked. This is a real danger. Just as we are beginning to research agroecosystems and how they work, we are tempted with more quick fixes, more opportunities to don the sorcerer's hat.

One such quick fix is biotechnology. Some of the most significant scientific advances in biotechnology are predicted for agriculture—indeed, biotechnology is being hailed as the next wave in agricultural progress—a quantum leap into a high-tech, genetically engineered future where all our problems will be solved. It is definitely dazzling in a gee-whiz kind of way—the way giant tractors dazzled my friends and me on the bus to school and the way DDT dazzled 1950s' farmers. It's so powerful, so technologically advanced, so promising, and how it works is so marvelously mysterious to the layperson. Because of all of the above, it is also easily hyped.

Keeping that in mind, I think biotechnology *could* potentially have a positive effect on pest management. Some possible positives include the creation of environmentally benign biopesticides and bio-fertilizers, and crops that have their natural resistance to pests enhanced. Perusing the latest agricultural journals, one finds several examples of this: a sprayable genetically engineered insect virus that gives newly hatched corn earworms a hormone imbalance and causes them to stop eating; a genetically altered variety of corn that produces an enzyme that curbs the appetite of a major corn pest; a natural insecticide in corn silk boosted through genetic manipulation, and so on.

These developments sound good, but I think we still need to be extremely cautious. Unintended side effects could emerge from such efforts.

Other biotech products, such as crops with built-in resistance to chemical pesticides, I suspect will not be sustainable in the long run. Much of the emphasis in biotech research has been on producing herbicide-resistant crop strains. Herbicide-resistant crops allow the routine use of a broad-spectrum plant poison while incurring no damage to the normally sensitive crop plant. 'Roundup Ready Soybeans' are an example, being immune to the effects of the popular herbicide Roundup. The proponents of herbicide-resistant crops argue that the approach is less costly, results in fewer pesticide applications, and actually uses less total herbicide than the conventional systems they replace.

On the negative side, exposed weeds will still continue to develop resistance to herbicides. Before long, some new genetically engineered species will have to be developed. It is important to point out that we have very little to go on when it comes to measuring the risks associated with releasing bioengineered life-forms. It is possible that genetically engineered releases, properly screened, could have very limited negative impacts. The fact is, however, no one knows for sure. Today biotechnology advances at a breakneck pace, faster than our ability to monitor its impacts on people and the environment.

And it is not enough for corporate agriculture to simply say "trust us." In recent years, the agricultural scientific community has had to admit that they too can be wrong—that the pesticides they claimed were virtually harmless can indeed be dangerous. Given this, a little humility about biotechnology is certainly in order.

Long Live the New King Cotton

> No plant has been the indirect cause of so much tragedy in the United States as has cotton. Slavery, the Civil War, soil exhaustion, extremes of poverty for millions, overcrowding of cities, racial troubles, all are legacies of the cotton crop.[48]

"The fleece of tiny lambs growing on trees" was how early Greek and Roman travelers described cotton. It is an essential crop: the most important fiber used to make clothing. Three out of four people in the world wear cotton clothing—from American T-shirts and jeans to Indian dhotis (loin cloths) and turbans.

(continued)

(continued)

Too bad its popularity does not make cotton easier to grow. Making cotton a sustainable crop will be a challenge: Myriad pests afflict it, so it's not surprising that cotton acreage in 1995 received more insecticide applications than corn, soybeans, or wheat. The boll weevil is still the cotton farmer's bugaboo, reportedly ruining about 20 percent of Oklahoma's cotton fields in recent years.

Perhaps some pest problems are due to the fact that two-thirds of the cotton in the major cotton-growing states was grown without any kind of rotation with other crops. Crop rotation not only builds the soil but can cut the incidence of plant diseases, insects, and weeds. The plant that held the imagination of my youth seems, unfortunately, to be a long way from being grown in a sustainable way.

At least that's what I thought until I discovered that cotton is being grown organically, most notably in Texas and California. A Texas A&M study confirmed that "organic production of cotton in West Texas was feasible and profitable."[49] For someone who grew up on a cotton farm in the 1950s, this seems fantastic in both senses of the word—both unreal and wonderful.

How do they do it? The Texas A&M study found that growing cotton in strips alongside forage sorghum was the "superior" production system. The two-crop system provided a refuge for beneficial insects, a crop to smother weeds, and residues to incorporate into the soil.[50] In California, big-eyed bugs, assassin bugs, and lacewings, all predators of cotton pests, were found to be more abundant in organic fields.

Researchers in the old Land of Cotton, the southeastern United States, have discovered that crimson clover and other legumes, into which cotton can be strip-tilled, are good winter and spring reservoirs for predators and parasites of cotton pests. Growing clover along field margins also helps provide refuge for "good bugs" as does leaving common weeds at field edges. Some cotton pests evidently prefer such weeds as fleabane and horsetail to cotton and these plants can act as "decoys" to coax bugs away.[51]

Of course, to realize these benefits, farmers have to stop their love affair with so-called "clean" fields and planting every square foot available to them.

Despite the challenges, more and more growers are going organic. In California, 10,000 acres were planted with organic or transitional cotton in 1994. Around Lubbock on the Texas High Plains, the Texas Organic Cotton Marketing Cooperative (TOCMC), formed in 1993, is going strong. This group of producers has gone from selling 400 certified organic bales in 1991 to 5,000 in 1999.

After overcoming big marketing hassles, TOCMC now sells their cotton to be made into a variety of fabrics. They also are developing their own products: blankets as well as facial pads, cotton balls, and tampons. These days, one can find all kinds of organic cotton clothing, as well as mattresses, stuffed animals, diapers, sheets, and towels.

And although consumers may like the idea of having chemical-free fabrics next to their skin, they can also feel good knowing they are contributing to the increased health of American agriculture. Organic cotton farmers fight weeds and insect pests using methods that are more complicated and often more expensive than methods used by conventional farmers. All are necessary if these pioneering Texas organic growers are to reach their stated goal of being "responsible stewards of the land in order for us to pass our farming heritage to our children. . . ."[52]

INTERACTIONS

Switching the emphasis of the Kerr Foundation's mission from conventional agriculture to sustainable agriculture in 1986 forced me to reexamine a belief I had held since I was a farm kid helping my family fight the boll weevil. I firmly believed it was impossible to farm without chemicals to control insects and weeds (or to farm without fertilizer). I grew up believing weeds were dangerous, among people who looked at them as a fault—a "good" farmer had a weed-free farm, period.

So what worried me the most after the organization's transition was the prospect of a farmer calling in need of advice about weed control and me not being able to respond with a specific chemical or a specific method of control. I didn't yet know much about alternatives—they seemed to me to be a closely guarded secret of pioneers in sustainable or organic agriculture. I worried that I would have to tell farmers that if they wanted to be sustainable they couldn't use chemicals ever—the equivalent of saying don't worry about the weeds, they'll go away.

Well, it was obvious to me saying *that* wouldn't impress your average Le Flore County rancher. So I had to go to school—independent study. I did my own research, walked through cotton fields in Louisiana and Mississippi, and talked to farmers who were leaving buffer strips of weeds and native grasses to harbor beneficial insects. Others I talked to were using crop rotation to control weeds. I soon realized that achieving a sustainable agriculture was a process—probably a long process—of change.

The first step in that process is understanding that sustainable agriculture is not farming-by-the-numbers. Making a farm sustainable requires more than knowledge of the proper poison to use to kill insects or weeds. It is a search for the root causes of a pest problem, and then working to solve those root problems. It is also tolerating some pests, including weeds. Gone are the days when we thought every weed had to go. It is too expensive and, in the end, unsustainable to kill every weed in sight.

So when I got back to the ranch after one of my forays out to find the true nature of sustainable agriculture, we began to do what we were asking farmers to do—change our thinking and our approach.

Our first challenge: sustainably managing the weeds in our own pastures.

. . . sustainable agriculture is not farming-by-the-numbers.

Over the years, we have experimented primarily with biological management of weeds. Our preferred tool is the four-legged weed eater, though we have employed the two-legged/two-winged as well. By using animals to control pests, we not only get safe pest control, but add diversity to our agroecosystems, both biological and economic.

The Kerr Ranch was once massive—60,000 acres of woods and pastures, a fitting kingdom for Senator Kerr, who had been dubbed the uncrowned King of the Senate in the 1950s. Today, the ranch consists of 2,500 acres of open pastures, still a very large holding in this part of Oklahoma. The ranch was managed in a conventional way until 1990; by 1993, the transition to sustainable management was mostly complete.

To make the ranch more sustainable, we wanted to optimize production and decrease purchased inputs, such as fertilizer and herbicide. To do this, we began rotational grazing of our cattle herds. Rotational grazing is the process of moving a herd from one pasture to another. Each pasture is grazed a limited number of days and then allowed a period of rest before it is grazed again.

The system has a number of advantages, one being weed control. Because the pastures, or paddocks as some call them, are relatively small and the number of cows confined to them relatively large, the cows are forced to change their foraging behavior. Generally, cows are selective in what they eat, avoiding weeds and seeking out young tender grass. In a rotational system, the cows' choices are limited and therefore they eat more or less what's before them, including some young weeds, which are more tender when young and therefore more palatable. Weeds when young are pretty easily digested and are moderately high in protein.

About the same time we instituted rotational grazing, we stopped applying commercial herbicides (and fertilizers). And although the rotational grazing has helped, weeds are still a problem. Why? The tame, improved grasses (mainly tall fescue and Bermuda grass) growing in the pasture are bred to respond to higher fertility than we have been supplying. So they don't compete as well with weeds as they once did.

To deal with the weeds in the short run, we are using mechanical controls: mowing pastures before weeds set seed. In the long run, such mowing is probably not sustainable because of the cost of fuel. But it is useful as the pastures change and forages that are better adapted to less inputs and local conditions gain over those less adapted.

What has happened to our pastures since herbicides and fertilizer supports were removed provides a classic dilemma. Tradition has been to maintain one forage type and graze the animals best adapted to it. But to preserve such monocultures requires large inputs of fertilizer and pesticides, because they are unnatural. Left alone, pastures will begin a succession and become more diverse—weeds and brush may increase in the short run. Now, diversity is good—but how to take advantage of it economically? To match the increased diversity of plant life in our pastures, we tried a diversity of animals.

Allelopathy

Some plants produce chemical substances (phytotoxins) that inhibit the germination or growth of nearby competing plants—a process called allelopathy. This phenomenon has been reported occasionally for weedy species of plants, but it also has been demonstrated in crop plants. How does it work? According to Harold Willis in his book *The Coming Revolution in Agriculture,* "Phytotoxins can be released from the roots of a growing plant or can be produced when plant residues decompose, . . . sometimes being released by certain soil microbes."[53]

Allelopathic plants can be used directly in crop rotations or as mulch, or the active substances can be extracted from the plant and used as herbicide. Such natural herbicides are generally broken down rather rapidly by common microorganisms and thus are not persistent pollutants in the environment, as are many of the synthetic herbicides. Rye, fescue, sunflowers, oats, and sorghum have been found to be allelopathic, some reportedly more effective than the widely used herbicide 2,4-D. Just recently, researchers have identified a substance in the bark of the tree-of-heaven *(Ailanthus altissima)* that can be used as a broad spectrum weed killer.

At the Kerr Center, we have conducted our own experiments with allelopathy. We collected leaves from common trees on our ranch such as black

(continued)

(continued)

locust, black walnut, hackberry, and water oak. We dried and ground the leaves of the different trees, then extracted solutions from each. We then tested to see the effects of various extractions (in water, acetone, alcohol) on the germination of wheat seeds and their early stages of growth using different cultivars. We also tested the extractions on radish and mustard.

We found that all of the leaves had allelopathic effects to some extent, some inhibiting germination, others inhibiting growth. We also found different plants more or less susceptible—radish seed germination, for example, was not affected at all by water oak leaf extracts while winter wheat and mustard were affected. We also found that what time of year the leaves were gathered made a difference. These experiments underscore how complicated nature can be. And how interesting.

The Cannon Multispecies Grazing project has been one of our more ecologically successful projects. Multispecies grazing refers to the use of more than one ruminant species (cattle, goats, or sheep) to graze a common forage resource, i.e., pasture or range. One species may follow another through a grazing area or two or more species may graze the same land at the same time. This practice has not been seen much on improved pasture—that is, pasture with nonnative domesticated species of grasses—but is more common on Western ranges where animals graze on native vegetation and the range is more diverse.

In the Cannon project, manager Elise Mitchell managed more than 100 Dorset-cross ewes on 180 acres. The sheep and cattle are rotated through pastures, sometimes grazing together, but usually apart. The diets of sheep and cattle overlap by only 25 to 35 percent, with sheep eating about twice as many forbs as cattle (forbs are broad-leaved nonwoody plants), including many weeds. As noted previously, cattle like weeds when they are young and tender, but sheep relish them at all times.

Before the sheep were introduced to one test area in 1988, ragweed, cocklebur, and pigweed were thick. By 1990, the weeds had been mostly wiped out by these four-legged brush hogs, so we expanded the forage area, and they continued to harvest weeds, while their companion heifers harvested grass. This project demonstrated that weeds are only weeds, as Mitchell likes to say, if there is nothing to eat them. What were once weeds are now forage for the sheep, who convert these resources into lamb and wool, salable products. We have gone from trying to maintain a monoculture forage base to encouraging species diversity.

We have also successfully used geese to eat crabgrass, Bermuda grass, and johnson grass in our organic strawberries. The use of geese for weed control is a good example of reclaiming an old practice that had been lost with the advent of the chemical age. Geese were once used extensively on mint and cotton fields to manage weeds.

We found that thirty-two Chinese weeder geese could control the grass in our two acres of strawberries. They were placed in the fields during the days after harvest in May and returned to sleeping sheds each night. Caring for the geese required only thirty minutes a day and the feed costs for thirty-two geese was forty dollars per acre. The geese were competitive with spraying herbicides and certainly safer. We especially wanted to demonstrate to organic growers that using geese could cut the time they spend weeding by hand.

Although farmers are supposed to look at things in a strictly practical dollars-and-cents kind of way, I would guess that even the most hard-boiled farmer could not help but be pleased by the sight of lambs in a green pasture or handsome white geese waddling through the strawberry patch. How much more satisfying, more earthy, more *right* it is than the sight of a man wearing a respirator with a sprayer in his hand and all it implies.

More and more people are beginning to understand that. Our experiences and the myriad research projects on alternatives to chemical pest control seem to beg the question: Are the solutions to our problems right at hand, on the farm, in the barns, fields, and fence rows, rather than in some chemistry lab at a major corporation in a far-off city? The prudent answer is that we are not there yet. Biological controls are not yet a viable and readily available alternative to the use of pesticides, and farmers often don't have the management skills to use these controls. That said, our work and the work of other researchers does show what can be done, and demonstrates what sustainable agriculture is all about. On one hand, it is a search for local solutions: working with what is at hand and what is adapted on the farm, instead of relying on costly inputs from off the farm. On the other hand, it is viewing possible solutions with a scientific eye, testing them for efficacy and economy. It is being creative and looking at the farm with an awareness of ecological interactions.

Sustainable agriculture is not farming-by-the-numbers, it is farming with an eye toward nature's cycles. Instituting a sustainable agriculture may require us to give up our ambition to take the place of the sorcerer. The sorcerer is Nature and we must accept that we will forever be the sorcerer's apprentice. Accepting our apprenticeship means we must keep learning. And learning about orchards, fields, and pastures as diverse agroecosystems is the first step to making an agriculture sustainable in the twenty-first century.

Integrated Pest Management: Many Little Hammers

In early summer, the cotton begins to bloom. In late June or early July, after the pale-yellow bloom fades and shrivels, there appears what is called a square. The square is three green sheaths, enclosing the cotton boll like a tent. The boll is the fruit of the plant that in three or four months will yield the cotton.

When the square appears, so does the boll weevil. Forty years ago, my father and I would walk down the rows and look for the telltale sign of the boll weevil—squares that had fallen on the ground. Then my father would decide whether it was worthwhile spraying the cotton. If the cotton was growing well and if there were a lot of squares, the crop was worth spraying. If there was a drought or a small number of squares, it wasn't worth the time and expense to begin the extensive spray schedule that cotton demanded. Making a good decision depended largely on the farmer's experience—we didn't count the squares and then plug numbers into a mathematical model to determine cost versus gain.

Because it was so subjective, we surely overused pesticides at times. On the other hand, we tended to be rather cautious in our use of chemicals because we were sharecroppers, without the means to be lavish. In this we were a bit out of sync with the times. Chemical pesticides in the 1950s were relatively cheap and the attitude was pretty much "the more the better," whether crops were suffering significant damage or not. Sometimes chemicals were sprayed to "prevent" any pests that just might be there from building up.

My father and I, like most farmers, had never heard the words integrated pest management, now known widely by the acronym IPM. But we were unwittingly following some of its precepts when we walked the cotton rows, looking for pest damage before deciding to spray. The term *integrated* was in the air in the 1950s, as high schools and lunch counters were racially integrated in Oklahoma and the nation. But integration in the context of integrating chemical and biological controls was largely unknown, though the term had been used in a 1959 article about control of the spotted alfalfa aphid. By the 1960s, it had become popular. Federal support for IPM began in 1972 and expanded in 1979.

(continued)

(continued)

What is IPM? A widely accepted definition is . . . "a pest management system that in the context of the associated environment and the population dynamics of the pest species, utilizes all suitable techniques and methods in as compatible a manner as possible and maintains the pest populations at levels below those causing economic injury."[54]

Although commonly thought of as an approach to insect management, IPM can be used for weed and pathogen management as well. According to agricultural economist Donald Vogelsang, IPM has the following three elements:

1. Diagnosis of the pest problem. Scouting or "field checking" is a popular means, but the term includes pest trapping and other methods.
2. Determination if and when a pest needs to be suppressed. Use of "economic thresholds" is most often used in making this decision. Generally, they are ratios such as numbers of insects or damage per 100 plants, above which the pest will cause an unacceptable yield loss. (For example, in the 1997 Cotton Insect Control Bulletin published by Oklahoma State University, farmers are advised to spray cotton for the boll weevil "when infested squares reach or exceed 15 to 25 percent," and keep spraying every three to five days until the infested squares drop below 15 percent.)
3. Suppression of pest. Any combination of several techniques can be coordinated to control a pest, including the judicious use of chemicals. The object of IPM is to hold pest populations below economically unacceptable damage levels. The object is not to eradicate them. Pest managers usually recommend one or more means of controlling a pest, but they let the farmer decide whether to act on the recommendations.[55]

This contrasts with the more conventional pest control approach in which pesticides are used routinely, often on a precise schedule that corresponds to the growth stages of a crop, in an attempt not simply to manage but to eradicate the pest. In IPM, mechanical, cultural, and biological methods along with chemical methods are used to keep pests under control. Some techniques include crop rotations, tillage, resistant cultivars, use of pheromones and other biological controls.

IPM can be spectacularly effective, cutting both pesticide use and thus cost to the farmer. In third world countries that were part of the 1960s green revolution in particular, there has been if not quite another revolution then at least a reformation with the enthusiastic adoption of IPM, in some cases as a national policy. Some farmers in China have reduced pest control costs by 85 percent. Brazil has seen an 80 to 90 percent reduction of pesticide applications in soybeans by introducing large-scale IPM programs. Taiwan's pesticide costs have

(continued)

(continued)

been reduced by 60 to 80 percent. India has cut rice and cotton pesticide use by large amounts while increasing yields.

Indonesia is perhaps the most impressive example of IPM's potential. In 1986, Indonesia's President Suharto banned fifty-six of fifty-seven pesticides because of resistance acquired by a major rice pest. No wonder: Some farmers were spraying their fields up to three times a week whether or not they had a pest problem. Since then, the Indonesians have instituted IPM on a large scale, relying on natural predators to combat pests while cutting pesticide applications and costs dramatically. At the same time, yields have increased.

These countries obviously have something to teach us this time around. IPM clearly has great potential in the United States, too. Some say it could reduce pesticide use by half. Some farmers are already reaping the benefits of IPM—reducing the number of sprays often with no decrease in the quality of the crop.

However, the way IPM is practiced in the United States has its critics. Some advocates of sustainable agriculture, such as Dennis Keeney of the Leopold Center for Sustainable Agriculture at Iowa State University, point out that IPM has been used only on a few crops, and has emphasized "appropriate use of pesticides rather than reducing pesticide use." He says that pesticide use has actually increased using IPM in many cases, and points out that it seems to have had little effect on pesticide use in Iowa, a key farm state.[56] As others have pointed out, American IPM programs all too often focus on pesticide management rather than pest management. A glance at university-produced fact sheets on pest control confirms this.

While IPM can have a real upside—sometimes reducing pesticide application and saving farmers money—it too often doesn't go far enough. Many are now calling for a form of IPM that relies much less or not at all on chemical pesticides. Called either biologically intensive pest management or ecologically based pest management, this new approach manages pests using natural processes supplemented by biological controls and pest tolerant plants, with specific pesticides targeted where needed.[57] Some have dubbed this new approach "many little hammers" because of its reliance on a variety of tactics to manage pests.

Checklist: How to Manage Pests with Minimal Environmental Impact

Weeds

Mechanical Approaches

1. Mowing
2. Flaming
3. Flooding
4. Tillage (including ridge tillage)
5. Controlled burns

Cultural Approaches

1. Crop rotation
2. Smother crops
3. Cover crops
4. Allelopathic plants
5. Plants spaced closely

Biological Approaches

1. Multispecies grazing
2. Rotational grazing

Chemical Approaches

1. Integrated pest management
2. Narrow spectrum, least toxic herbicides
3. Properly calibrated sprayers
4. Application methods that minimize amount used, drift, and farmer contact

Insects and Diseases

1. Enhance existing populations or introduce natural predators, pathogens, sterile insects, and other biological control agents
2. Traps
3. Maintain wild areas or areas planted with species attractive to beneficial insects
4. Trap crops
5. Selective insecticides or botanical insecticides that are less toxic
6. Cover crops
7. Crop rotation (avoid monoculture)
8. Intercropping, strip cropping
9. Maintain healthy soil (prevents soil-based diseases)
10. Keep plants from stress

Chapter 8

Power: Step 7—Conserve Nonrenewable Energy Resources

> Sustainable approaches [to farming] are those that are the least toxic and least energy intensive, and yet maintain productivity and profitability.
>
> from the University of California
> Sustainable Agriculture Research and Education Program
> *What Is Sustainable Agriculture?*[1]

Oklahoma has a special relationship with the fossil fuels—oil, gas, and coal. The production of fossil fuels has been a big part of the state's economy. Early in the twentieth century, oil wells sprouted like wildflowers across the Oklahoma prairie. Oklahoma entered the Union in 1907, leading the nation in oil production; the state currently ranks fifth. Oklahoma has been home to some of the country's biggest oil companies—Phillips Petroleum (of Phillips 66 fame), Kerr-McGee, Sinclair, Continental (now Conoco), and Getty. The state's ties to the oil business are visible in Oklahoma City, where oil wells pump on the grounds of the Capitol.

It's not uncommon, as you drive across the state, to see cattle grazing next to an oil well. Many farms have had oil or gas wells on them, making some farmers rich or at least providing a small royalty check from time to time. Because Oklahoma is such a young state, the agriculture and petroleum industries have developed side by side. Both have seen extravagant booms and equally intense busts; the two are entwined in a profound way.

In Oklahoma, the cattle came first. The great cattle trails of the late nineteenth century—the Shawnee, the Chisholm, the Goodnight-

Loving—crossed Oklahoma. Cattle were driven north from Texas across Oklahoma to railheads in Kansas and on to the growing cities of Gilded Age America that were clamoring for beef.

The sodbusters came later, most dramatically in the land runs into Oklahoma Territory of 1889 and 1893. The area around my hometown of Cold Springs was opened up to non–Native Americans in 1901. Land was taken from the Kiowa, Comanche, and Apache, who had themselves been resettled there not long before; this same land was distributed to new settlers by lottery. By then, the great trail drives were over, the drovers and dogies settled down, and cattle ranching and farming began in earnest.

At about the same time, oil exploration began, particularly in the eastern and central parts of what would be Oklahoma. At about the same time that Cold Springs was coming into existence as a farming community, the Phillips brothers hit the first of what would be eighty-one producing wells in a row. Such luck made ordinary men into millionaires nearly overnight. One of these was Robert S. Kerr, who was born in a log cabin on a small farm in southern Oklahoma. A small part of the Kerr oil fortune, in trust, funds the Kerr Center.

As Kerr was well aware, oil makes America run. Oil is the fluid that lubricates and powers our cars, and one might say, makes our lives slicker and easier. The great thing about fossil-fuel energy is that it is concentrated and its energy is released so easily. The bad thing is that, unlike the solar energy which created it, it is finite, and we are using it up fast. In 1890, total world energy use was one terrawatt (equal to five billion barrels of oil per year). In 1990, world use was nearly fourteen terrawatts.[2] The increase is due both to increases in world population and in energy use per person.

Senator Kerr died in 1963 during the glory days of Big Oil and cheap energy for the American consumer. By 1982 in Oklahoma, the oil business was still in high gear—a record 12,030 oil wells were completed that year. By 1996, though, total well completions in Oklahoma had dropped to 1,607.[3] Though low prices have certainly contributed to this decline in drilling, the truth is, Oklahoma's oil is running out. What has happened in Oklahoma will one day, in the not-too-distant future, happen worldwide. The well will run dry.

Despite this certain future, the United States is still consuming more energy than it produces. We consume about 25 percent of the world's energy and produce about 20 percent. Oil is not sold as if it were a finite resource, and while prices remain low there is little incentive for conservation. Low prices have lulled us into complacency. Things will change: I predict that price increases of over 10 percent will force people, including large numbers of farmers, to begin looking at new ways to seriously reduce energy use.

Changes may not be long in coming. Although industrialized countries such as the United States consume about ten times the amount of energy per capita as developing countries, in some developing countries fuel use is increasing, accounting for two-thirds of the rise in demand between 1985 and 1995,[4] and thus adding to pressures on world supply. For example, since 1985, energy use is up about 30 percent in Latin America, 40 percent in Africa, and 50 percent in Asia. The Energy Information Administration forecasts that worldwide demand for oil will increase 60 percent by 2020.[5]

Because oil-supplying nations are pumping oil at a rapid rate, prices have remained low. Eventually, however, as demand increases and supply dwindles, the prices will inevitably rise. This will make it more profitable to get sources of oil, gas, and coal that are more difficult to retrieve. However, some experts believe that the discovery of a few more large oil fields will make little difference against consumption on today's scale.[6]

Indeed, long-time exploration geologists and energy analysts Colin J. Campbell and Jean H. Laherrere, writing in *Scientific American,* have ascertained that the earth's conventional crude oil is almost half gone. They assert that oil industry estimates of reserves are exaggerated, that the discovery of new oil is not close to keeping up with the amount being pumped, and that within the next decade the supply of conventional oil will be unable to keep up with demand. From an economic perspective, they write, when the world runs out of oil is not directly relevant: what matters is when production begins to taper off. This, they predict, will occur before 2010.[7]

The upshot is that unless demand declines, prices will rise. And if these analysts are to be believed, it will happen soon—the era of abundant, cheap oil is about to end. Petroleum may be scarce by the middle of the new century, and natural gas may not last much longer.

Some put their faith in some unknown miracle technology that will save us before our lives and industries are disrupted by fossil-fuel scarcity. The situation seems to be rather that no viable alternatives to fossil fuel are even close to being online. Harnessing tidal power poses locational problems, because few cities are located near big tidal flows. Fusion power remains beyond reach, and there are environmental problems with hydroelectric projects.[8]

Hydrogen is a possible substitute for fossil fuels. Currently, the hydrogen to power prototype cars is obtained from natural gas or methanol. In theory, hydrogen could be produced from water by solar-power converters, but such technology is not yet practical.[9] Whatever alternatives are found, the changeover from fossil fuels to some other energy would likely require massive amounts of capital.[10] It is likely to be a rough transition for everyone once the finite nature of fossil fuel-energy finally becomes a reality.

AGRICULTURE'S ACHILLES' HEEL

Chapter 1 called the story of agriculture in the twentieth century a tragedy. In a classic tragedy, the hero has a fatal flaw, one that ultimately brings him or her down. Achilles was invulnerable everywhere except on one heel, where he was vulnerable to the arrows of his enemies. One fateful day, an arrow found his heel and laid him low. Industrial agriculture's dependence on fossil fuel is its fatal flaw, its Achilles' heel.

In this section of the book, we explore energy and how intertwined it is with money in agriculture. Which road we take to the future of agriculture will be largely decided by what vehicle we choose to drive—one powered by fossil fuels and only affordable to the few, or one powered by the sun and affordable to many. The vehicle we choose will be either industrial agriculture or sustainable agriculture.

It used to be that farms were fueled primarily by human labor. As Robert Frost put it in his poem "Mowing":

> There was never a sound beside the wood but one,
> And that was my long scythe whispering to the ground.[11]

By 1890, horse-powered machinery had transformed American farming. In the 1930s, the all-purpose rubber-tired tractor with its implements had come into general use, transforming agriculture again. It wasn't until 1955, though, that tractors outnumbered horses and mules on American farms.

Today, it's hard to imagine that. Ads in farm magazines trumpet the latest—at a local farm implements dealer I can buy a "Power Plus 8570" tractor for a mere $83,137. Today the silence in the fields that Frost wrote of has been drowned out by the sputter and roar of huge diesel-powered machines.

What has clearly changed in agriculture over the past couple hundred years, with the change accelerating after World War II, is that human (and animal) energy has been replaced by mechanical power. This mechanical power is fueled by petroleum energy. Although the use of machinery is more economical than the use of labor, the cost of operating and maintaining such machinery still forms a large part of the farm budget. As I was watching agriculture change in Cold Springs during the 1960s, such costs went up by about one-quarter. Not surprisingly, the power used, calculated in horsepower, also went up, by about one-third. Man hours during the same period went down.[12]

With Oklahoma having produced over thirteen billion barrels of oil and sixty-nine trillion cubic feet of natural gas, the state has certainly done its part to keep the fertilizer flowing and the tractors of America running. Without a doubt, fossil fuels make industrial agriculture run. As one-time USDA secretary Earl Butz put it: "U.S. agriculture is the number one customer of the petroleum industry."[13]

Agriculture and Energy: The Numbers

A number of ways are available to gauge the relationship between industrial agriculture and energy use. One is the fact that farm use amounts to about 23 percent of the nation's energy, comparable to the mining and construction industries.[14] Another is that industrial agriculture is energy inefficient compared to other types of agriculture. Industrial agriculture as practiced in the United States uses about twelve times as much energy as agriculture in less-developed countries of the world.[15]

To produce some of our most important food items, we invest much more caloric energy in the form of tractor fuel, fertilizer, drying, processing, and shipping than these foods give us in calories. Among these: feedlot beef with ten calories invested per one calorie of food produced, and intensive poultry production with a ratio of two to one. Compare this to milk from grass-fed cows at one calorie in, one out, and range-fed beef at half a calorie in for one calorie's yield. Far more efficient than these, however, is traditional wet rice culture with 5/100 of a calorie invested for every calorie yielded.[16]

Another way to look at it is that although total and per-acre production of food in the United States vastly increased with the advent of industrial agriculture, output per unit of energy actually declined. For example, in the glory days of the 1970s, farmers used 50 percent more energy to grow 30 percent more crops.[17] The energy inefficiency of mechanized agriculture, it has been pointed out, is unimportant as long as fuels are cheap and plentiful, but it is worrisome as fuels become scarce.[18]

To produce some of our most important food items, we invest much more caloric energy in the form of tractor fuel, fertilizer, drying, processing, and shipping than these foods give us in calories.

One reason for this inefficiency is that foods that were once grown, processed, and eaten at home on the farm, are now subjected to a processing, packaging, and distribution system that is energy intensive. About three to five times as much energy is consumed in these off-farm activities as is used on the nation's farms. This is largely due to regional specialization of food production. Because of this regional specialization, food must be shipped farther, and also much of it must be processed (cooked, frozen, canned, etc.) to avoid spoilage on the long trip from farmer to consumer. Estimates range as high as 1,250 miles between farmer and consumer for the average food item.[19]

This system has not directly benefited the farmer—a farmer receives, for example, only a few cents for the wheat in each loaf of bread that is made, and yet the grocery shopper often pays more than a dollar. Because of this, an agriculture that is sustainable in energy-use must modify the extreme regional specialization that is currently the norm, diversify local crops, and focus more on local markets. These changes are already occurring on a small scale, as evidenced by the growing popularity of farmers' markets, subscription farming, and other direct farmer-to-consumer arrangements.

But what about the energy used in food production on a typical industrialized farm itself? If you flew over rural America in a small plane on any given day at planting or harvesting times, you would see farm machines, powered by internal combustion engines, crawling across the fields, burning fossil fuels. Large amounts of energy are used directly on the farm—in fuel that goes into tractors, trucks, and other machines for all the ordinary jobs on the farm—tillage, planting, cultivating, harvesting, and then transporting the harvested crop to market.

Other uses, though intensive, are less direct. Many agricultural chemicals are derived from fossil fuels. Then there is the fuel needed to manufacture fertilizer and chemicals; the fuel needed to distribute them; the fuel needed to mine and distribute lime, as well as phosphates and potassium, two other key elements in fertilizer; the fuel needed to make tractors and farm implements. Other uses: electricity (often from gas or coal-fired plants) used to power irrigation pumps and other farm machinery such as dryers.

The Farmer's Energy Costs

It is the field-level view, however, that absorbs the farmer's attention. One way to gauge the extent of fossil fuel use is by looking at a sample farm budget. In 1997, a farmer raising corn in the north central part of the country could expect to spend 62 percent of total variable cash expenses for manufactured inputs, either fossil-fuel based or using fossil fuels to be mined and delivered. This breaks down to 36 percent for fertilizer, lime, and gypsum; 19 percent for chemicals, and 7.6 percent for fuel, lubrication, and electricity. Out of the crop's total expenses (variable and fixed), these items add up to 48 percent.

Large amounts of energy are used directly on the farm—in fuel that goes into tractors, trucks, and other machines for all the ordinary jobs on the farm—tillage, planting, cultivating, harvesting, and then transporting the harvested crop to market.

A soybean farmer in the same part of the country would spend slightly less on these manufactured inputs for of his variable cash expenses—56.3 percent. He would spend less for fertilizer, lime, and gypsum (soybeans are a legume and make nitrogen)—12.3 percent; but more for chemicals—33 percent; and slightly more for fuel, lubrication, and electricity—11 percent. Here, too, energy-related expenses are a sizable portion of his total expenses—34 percent.[20]

Expenses are similar in Oklahoma. A bottomland soybean farmer spends 22 percent of his total operating costs on herbicide alone.[21] Nationally, farmers spent 23.4 billion dollars on such manufactured inputs in 1995.[22]

In these days of low prices, it's easy to see why a farmer would want to cut these costs. It is also easy to see how step 8—increase profitability and reduce risk, covered in the next chapter—cannot be accomplished in the long term without step 7—conserve nonrenewable energy resources—the subject of this chapter.

LOW-INPUT FARMING

Both in the short and long run, reducing energy costs will allow producers to save money. Reducing dependence on oil will help the farmer stabilize cash flow as energy prices rise or supplies become limited.[23] Currently, though energy prices are still low, so are profit margins, and saving money is a compelling reason to cut energy inputs for many conventional farmers.

This brings me back to April 1988, when I testified before Oklahoma Representative Synar's Environment, Energy, and Natural Resources Subcommittee. It has been true for a long time that reducing input costs is key to the American farmer's ability to stay in business. As I told the committee then, ". . . a low-input agriculture system could have helped avert the mass exodus of people off of the farm. This is because a more natural system of production would yield less per unit of land, but could keep that yield at a more sustainable level over time. This would be helpful in controlling the volatility of farm commodity quantities and reduce the boom-and-bust cycles of production."[24]

Another witness that day was Carl Pulvermacher, a dairy farmer from Wisconsin, who had used few pesticides or chemical fertilizers on his farm since 1982. He said: "Our system has the added benefit of reducing our dependence on petroleum and petroleum-based fertilizers and pesticides. Because we don't depend on the petroleum-based products, we aren't as responsive to changes in the petroleum industry as conventional farmers are. Our actual fuel use is typical—all of our practices add up to four trips through the field each year. This compares with my conventional farmer neighbor who makes four trips to apply NH_3, bulk fertilizer, and pesticides for one cultivation."

Pulvermacher also compared his expenses for pesticides and fertilizer ($200 in 1987) to his neighbors ($6,000 to $10,000 annually). He commented that his thrifty approach was the main reason he was still in business. Furthermore, his yields of milk and corn have been higher than the state average. In fact, he even won a farm production contest sponsored by the Wisconsin Soybean and Corn Growers Association and the University of Wisconsin Extension Service for having high yields and low production costs![25]

In recent years, even farmers who have more or less stuck to a conventional system have realized that one way to save money on fuel bills is to reduce the number of "trips" they make across the field, i.e., how often they go into the field to till the soil. Many farmers are now using reduced tillage or conservation tillage techniques to do just that. I like the term "residue farming" to describe these methods, because they leave at least 30 percent of the soil surface covered by crop residue.

. . . his thrifty approach was the main reason he was still in business.

Farmers using conventional tillage spend a lot of time preparing their fields for planting. They might plow, chisel, disc, or harrow, operations designed to loosen the topsoil and eventually pulverize any big clods of dirt, leaving the field relatively smooth, perfect, as the thinking goes, for planting seeds. Getting this may require many trips across the field on the tractor pulling various implements, particularly in fields in which the soil has become hard and compacted from the weight of big tractors rolling across them, and where the soil has little organic matter from years of industrial farming, and the texture is therefore dense and heavy.

Ridge Tillage

The Kerr Center has worked with an Oklahoma soybean farmer, Kenneth Repogle, who a few years ago decided that chiseling his 1,200 acres twice in a season was largely a waste of time and diesel fuel—i.e., money. He then decided to try ridge tillage, which, to my mind, is the most promising of the reduced tillage methods. His experience proves that farmers can save energy on the farm by improving energy efficiency without reducing their productivity.

In ridge tillage, the soil is not tilled between harvest and the next planting. The crops are planted in a seed bed on permanent ridges instead of on flat ground. Special implements are used to heap up the surface soil in a series of what are essentially raised beds at regular intervals. After the crop is harvested, residues are left on top of the ridges. When it comes time to plant again, ridge-till planting attachments scrape what crop residues remain off the top of the row ahead of the planter. (Repogle runs a harrow [like a big rake] over the ridges before planting to knock down stalks.)

Ridge tillage has a number of advantages—drastically improving soil quality, for one. Although the soil on neighboring farms is hard, Repogle's soil is soft and friable. This is because of the increased organic matter in the top layer of the soil. After Repogle harvests his crops, he leaves the remains—roots, stems, leaves—on top of the ground instead of plowing them under. The residue as it decays forms mulch that protects the soil from driving rain and keeps it soft. The decaying residue adds organic matter to the top inches of the soil. The effects of this residue, along with never driving on the ridges with heavy tractors, keeps the soil soft and not compacted, allowing plant roots freer growth. The soft soil also absorbs water, halting excessive runoff. Earthworms have returned to his fields, a sign of the improved health of the soil.[26]

The change in water absorption has perhaps been most dramatic. Repogle used to chisel his ground twice before planting (a chisel is a clawlike implement that rips the soil open, allowing for greater water absorption). Chiseling, because of the power it takes to rip the soil, is an energy-intensive operation. Despite these efforts, after heavy rains one could look over his 1,200 acres of bottomland and see a lake of water.

Since switching to ridge tillage, Repogle reports that the water is absorbed quickly—no more lakes, despite the fact that he farms lowland between two rivers. Soil erosion is virtually nil; his drainage ditches, once clogged, are clean of sediment. And because of his soil's increased moisture retention, his beans do better in droughts. In fact, one of the main pluses of ridge tillage to him is the way that it evens out his production from year to year—in bad weather years, he does better than neighbors using conventional tillage and in good years, he equals them. He is amused by the fact that some of his neighbors still seem puzzled by the difference in moisture between his place and theirs.

And what about energy use? Repogle says he gets "as good or better bean yields with about three or four fewer trips across his field," a savings of three gallons of diesel per acre.[27] His herbicide costs have also dropped; the ridge cultivator he runs twice after the plants emerge cuts down weeds between the ridges, and smothers weeds near the plants when it throws dirt up to build the ridges. The herbicide that he does use he bands (applies just next to the row). By doing so, his use has dropped by one-third.[28]

Other types of conservation tillage also save in direct fuel use; some of these savings are dramatic. Fuel costs in no-till systems can be less than 50 percent of those in the moldboard tillage system for corn, soybeans, and winter wheat. However, in no-till, herbicides are used heavily and therefore, from a macro viewpoint, the energy savings may be somewhat less, as these herbicides are made from fossil fuels, and fossil fuels are expended in their manufacture and distribution. So although no-till is decreasing soil erosion and direct fuel costs, it doesn't appear to cut indirect fuel costs in the form of herbicides.

Repogle says he gets "as good or better bean yields with about three or four fewer trips across his field," a savings of three gallons of diesel per acre.

You would think that Repogle's success would be copied far and wide, but not as of yet. A few neighbors have emulated him, and he believes change is coming, albeit very gradually. As he put it when the President's Council for Sustainable Development visited his farm in September 1997: "Farmers are hardheaded suckers."[29]

What are the reasons for this hardheaded attitude? Energy experts Amory and L. Hunter Lovins and Marty Bender have identified four reasons farmers continue to depend on fossil fuels. One is the capital intensity of modern farming. After investing large amounts of money in conventional equipment, it can be financially difficult to buy new equipment such as a ridge cultivator. Repogle had trouble getting a loan for equipment because at the time Oklahomans were unfamiliar with the technology. Another reason is the lag time in perceiving that a formerly successful process is yielding diminishing returns. At the core of the stubbornness Repogle refers to is probably this mind-set—the reluctance to accept that the old way is not working.

In addition, all the pressures from the outside weigh heavily—the agribusiness lobby, say the Lovinses and Bender, has a vested interested in keeping American agriculture hooked on oil. Ads taken out in farm magazines and money to researchers in universities from these businesses reinforce dependence on fossil fuels. It is the status quo, as expressed in Indictment II, supported by the USDA, extension, and agriculture educators in the universities.

The last but not least barrier to change is the lack of alternatives.[30] Sometimes one has to search far and wide for new, more energy-efficient ideas. Kenneth Repogle read of success with ridge-till in other parts of the country, but could find no information locally. He had to make a trip to Iowa and three trips to Missouri to get the information he needed.[31]

One hopes at least some of these barriers to change will collapse, hopefully before the predicted oil crisis hits. That is why research into sustainable agriculture methods is so important. Even within a conventional system, a number of ways exist to decrease fossil-fuel inputs. Banding pesticides and fertilizers, soil testing before applying fertilizers, and making sure that sprayers are modulated to put out the proper amount are common sense, conservative approaches to saving on high-energy inputs.

It is also important to keep in mind optimum returns from fertilizer use. As it turns out, adding more and more fertilizer does not result in more and more yield. Whereas putting on a certain base amount raises yield dramatically, doubling that amount will not double yields, but instead raise them by a small percentage. A point exists where extra application does not pay; every farmer should know that point.

Such measures tweak the industrial system in the direction of energy conservation. But for a farm to be energy sustainable in the long run, it must replace nonrenewable fossil-fuel energy as much as possible with renewable solar energy. This often means replacing off-farm inputs with on-farm inputs and management techniques that are more energy efficient.

Saving Energy with Crop Rotation

In a 1995 article in the journal *Sustainable Farming,* six typical cash-crop farms in Quebec were examined. For each farm, the researchers figured the ratio of energy output to input. For energy input, they counted the energy required to produce, transport, and use on-farm inputs such as fertilizers, herbicides, fuel machinery, etc. They found that one farm, labeled Farm 1, had an energy efficiency of twice that of another farm, labeled Farm 2.

Why? Both farms had corn as their most important crop. The most important energy input in growing corn is nitrogen, accounting for 39 percent of total energy input on these farms. The energy input is the energy that is required to produce, transport, and use the various inputs. Other energy inputs, in order of descending importance, were corn drying (37 percent), fuel and gas (18 percent), machinery (5 percent), and herbicides (2 percent). What the researchers found was the more efficient Farm 1 had found a way to reduce inputs, particularly nitrogen.

Farm 1 accomplished this through crop rotation. The rotation began with corn, was followed by soybeans, and then a cereal crop sowed with a clover green manure, then corn again, renewing the cycle. Because the soybeans and the clovers are legumes that fix nitrogen, some nitrogen remains in the system and can be used by the corn and the cereals, reducing the need for off-farm purchase of synthetic nitrogen. This farm further saved energy by sun drying at least half of its corn in a traditional corn crib, rather than drying with propane or electricity (see Figure 8.1).

FIGURE 8.1. Agricultural Ecosystem

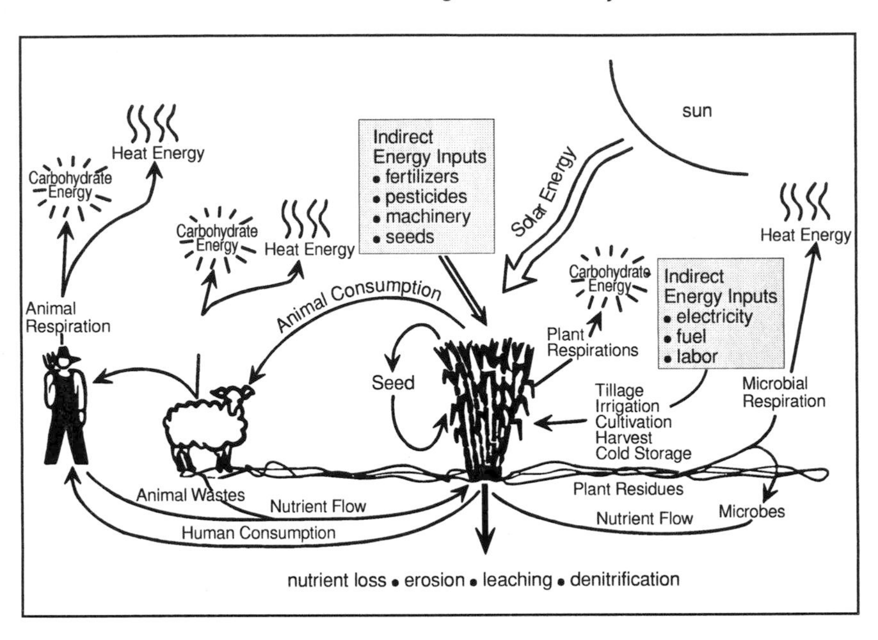

Source: USDA/NRCS WSSI—Sustainability Technical Note 1, October 1997.

Organic systems, which rely heavily on crop rotation to make the system work, can save energy. In the pamphlet "Switching to a Sustainable System," North Dakota farmer Frederick Kirschenmann presents a number of model farming systems. Their common denominators: rotations that fix nitrogen; rotations which incorporate green manures and addition of livestock manure; weeds controlled mechanically or through allelopathy, rather than chemically.

In contrast to no-till systems, the reviewed systems relied heavily on tillage. But, Kirschenmann asserts, that does not necessarily mean higher fuel costs. On his own operation, his fuel costs have gone down due to two factors. First, because he has built up the quality of the soil and thus the tilth, less horsepower is needed to pull tillage equipment. Second, he relies more on shallow, minimum tillage operations that require little horsepower. Moldboard plows are used rarely. Deep-rooted plants in the rotation open the subsoil.[32]

Conserving energy on the farm inevitably comes back to how well the farm can imitate natural systems. In a natural ecosystem, solar energy flows into the system and powers the growth of plants. When the leaves fall or the plant dies, this organic matter as it decomposes slowly returns to the soil most of the nutrients that had been taken up by the plant. It is a closed system, a closed nutrient cycle, nature's ultraefficient recycling program.

In an agroecosystem, solar energy also flows in and powers the growth of plants. But the farm system is not closed. Much of the crop is taken out of the system—in cereal crops, a third of the biomass; in corn crops, half. Therefore, a large portion of the nutrients are carried away and are usually replaced by fossil-fuel inputs in the form of fertilizers.[33]

Conserving energy on the farm inevitably comes back to how well the farm can imitate natural systems.

Crop rotation mimics a natural system in a couple of ways. In a natural system there is a diversity of plants. This is also true in systems that use rotations—different crops take out and contribute different nutrients to the system. Legumes are an excellent way to add nitrogen and therefore are important plants in natural systems.

Other new farming practices also mimic aspects of natural systems. In ridge-till and other conservation tillage systems crop residues are left on top of the ground such as in a natural system, rather than being burned or plowed under. This contributes organic matter and nutrients to the topsoil. The slow release of other on-farm inputs such as green manures and organic fertilizers (such as manures and compost) mimic the release of nutrients in a natural system. The end result of these approaches: an energy-efficient, energy-conserving farm.

Checklist for Farmers:
How to Conserve Nonrenewable Energy Resources

1. Reduce number of tillage operations.
2. Cut use of chemicals and fertilizers.
3. Develop production methods that reduce horsepower needs.

(continued)

(continued)

4. Recycle used oil.
5. Use solar power when possible.
6. Use renewable, farm-produced fuels: ethanol, methanol, fuel oils from oil seed crops, methane from manures and crop wastes.
7. Use crop rotation to add nitrogen.

Solar Water Pumping

Solar energy offers a readily available and practical energy source to meet a variety of needs on the farm, including charging electric fences and pumping water. Depending on the application, water pumps may be powered by solar energy using photovoltaic cells. Traditional pumping systems, which employ diesel engines and electric grid-powered motors, represent a partial solution for some water delivery needs. However, the cost of fuel and electricity, spare parts and service—or the equivalent in time and labor of hand pumping systems—often make pumping water extremely expensive.

Photovoltaics is the science of using the sun's energy to create electricity. When sunlight travels to the earth, it is in the form of invisible energy units called photons. When these photons hit photovoltaic cells (made of silicon) on a solar panel, the light energy from the photon is transferred to the silicon and a usable electric current is generated.

Photovoltaics are extremely reliable—the cells have no moving parts to wear out, require little maintenance, and last virtually forever. Photovoltaic power can meet a variety of pumping requirements, from a low-horsepower shallow-well pump to a high-horsepower irrigation system.

Solar-powered water pumping can be very useful in agriculture, since crops generally require the most water during the sunniest times of the year. Solar power can also supply water for livestock, which can increase the productivity of grazing operations. The high reliability of solar cells allows for unattended operation in remote pasture areas and fields.

Figure 8.2 shows a portable, solar-powered pumping system that was installed at Meadowcreek, a nonprofit environmental education center in Arkansas that was affiliated with the Kerr Center for several years. It is designed for use in shallow ponds and streams. Because it is portable, the Meadowcreek pumping system can be used wherever it is needed for irrigation, stock watering, pond aeration, or other farm, gardening, and agroforestry uses. It uses twelve photovoltaic modules to produce electricity, which powers a direct current centrifugal jet pump. Under full sun, this system produces 720 watts and delivers 13.8 gallons of water per minute! The system is capable of pumping over 4,000 gallons per day from a depth of sixty-six feet.[34]

FIGURE 8.2. Diagram of Solar-Powered Water Pump

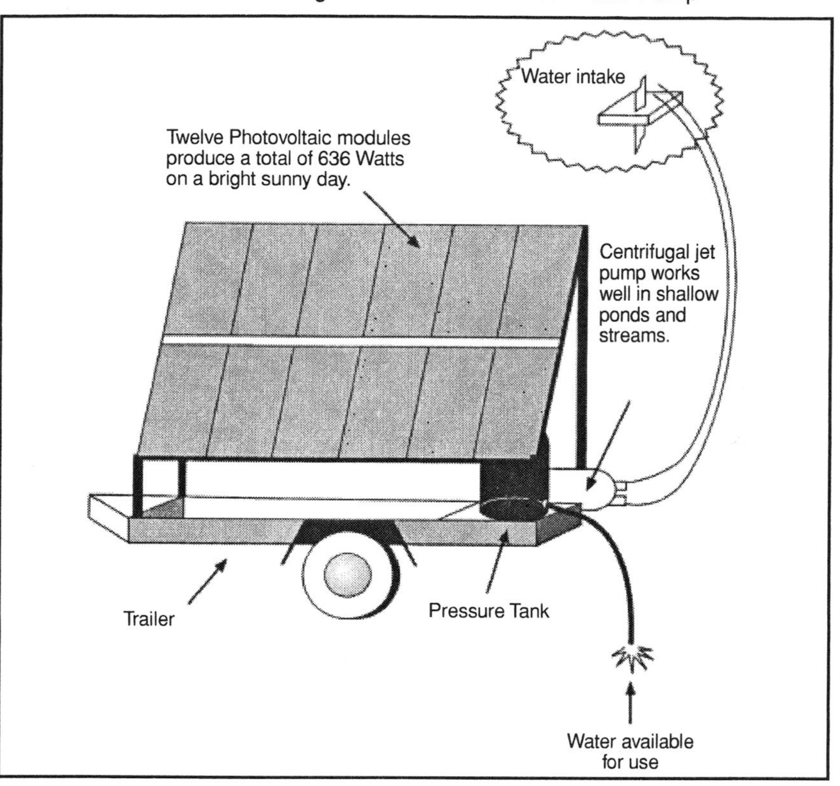

Source: Meadowcreek (Kerr Center) Technical Brief No. 2 (1992).

Chapter 9

Money: Step 8—Increase Profitability and Reduce Risk

Our 95 acres is paying three full-time salaries. The average farm in America takes $4 of machinery and buildings to turn one dollar in annual gross sales. Our ratio is 50 cents to a dollar. However you slice it, however you want to cut it, we can make the numbers work. The beauty of this is when we farm ecologically, we also farm economically. That's the beauty of this. We don't have to sacrifice the one to get the other.

Shenandoah Valley farmer Joel Salatin[1]

Economic activity must account for the environmental costs of production. . . . The market has not even begun to be mobilized to preserve the environment; as a consequence an increasing amount of the "wealth" we create is in a sense stolen from our descendants.

William D. Ruckelshaus[2]

Power and money; money and power. Industrial agriculture runs on both. Power or, more specifically, massive infusions of fossil fuels, makes industrial agriculture run. Money or, more specifically, massive infusions of capital, is also essential. If power is the gasoline, then money is the oxygen—and the mix revs the engine of industrial agriculture. Unfortunately, industrial agriculture has made the full-time small farmer, Jefferson's ideal citizen, as rare as a Cadillac on an Oklahoma country road.

Driving from the Kerr Center to my home twelve miles away, I pass by a number of farms. Few are operated by a full-time farmer who is supporting a family. Who are my neighbors? Some are retirees from

California who raise cattle for extra income. Others are people like my wife and me who are professionals and raise cattle on the side. There are elderly retired farmers—couples who did raise families on small farms in the little green hollows that make this area so picturesque. They may still run a few cattle, too, and perhaps grow a garden and a few peaches and pears in the family orchard.

Things are changing in Le Flore County; agriculture is changing and so is land use. More and more often, when these retired farmers die, their children, who now live in far-off cities, sell the home place, which often ranges from forty to 160 acres, to developers who subdivide it into five- or ten-acre plots. Then those who want to retire to the country or those who work in the city of Fort Smith, Arkansas, twenty-five miles away, move in, often in mobile homes. Le Flore County has become part of the metropolitan-influenced area of Fort Smith, transformed into a spacious, rather distant suburb, but still city centered in its economics.

I confess that I hate to see these farms divided up. It goes against the grain in a deep way. Every time a farm disappears, I see a little piece of a whole way of life—a valuable way of life—disappear. However, there are still some who are trying to make it in farming. In 1992, 676 families listed farming as their principal occupation and had at least $10,000 in gross sales (see Figure 9.1). (The population of Le Flore County is around 45,000.) And this number was up sharply over the 1982 figures, due to the expansion of contract chicken and hog production in the area. Contract farmers assume gigantic debts—around one-quarter of a million dollars to build a chicken house to company specs, in order to earn low wages.

The downside is that even for farmers who consider farming to be their principal source of income, it is unlikely that their farming enterprises are paying all of the family's expenses. And Le Flore County continues to be a poverty pocket, with per capita income running at approximately 78 percent of the state average.

For the vast majority of American family farmers, farming is a part-time, minimally profitable, or downright unprofitable enterprise. Farm income accounts, on average, for only 12 percent of farm household income; contrast that to 46 percent from wages and salaries.[3]

The USDA classifies farms in the United States according to their gross sales. "Noncommercial" farms are those with less than $50,000 in gross sales. "Commercial" farms have at least $50,000 and are further divided into categories from small ($50,000 to $99,000 in gross sales) up to superlarge (one million dollars or more in sales). For the 1,514,476 noncommercial farms existing in 1993, nearly 75 percent of all farms, net farm income averaged $1,105.[4] Even among the "commercial" farms, those that are supposed to be economically viable, half of total income came from sources off the farm.

Supporting a household at a good standard of living through farming enterprises seems to be impossible for most farming families today. Only 8 percent of all farm operator households received income from farming near or above the average income for all U.S. households.[5]

So the experience of my wife and me and our two children is perhaps pretty typical. We both work full time (my wife Brenda teaches)—and the first 160 acres and our home were financed by those jobs. We borrowed more money and slowly built a cattle herd; on weekends we built corrals and fences while our children were growing up. Brenda says that the next eighty acres we bought on her salary. Every year we try to find some money to invest back into the farm. Last year, we cleaned out a pond; this year we want to build a pond in another pasture, and after that we need to lime the pasture. The list goes on. We've planted trees in the pastures. I've applied chicken litter in an attempt to improve the fertility of the soil.

We always try to make enough from the cattle to pay expenses and finance capital improvements. If cattle prices are low, we dip into our salaries. The economist in me wonders whether, say, spending forty dollars per acre to add limestone to make nutrients more readily available to the grass is worth it. Cattle simply aren't that profitable. Or, when I am pressed for time, I can think of things I would rather do than fix a barbed-wire fence on a Saturday afternoon. But good fences make good neighbors. And since my Hereford bull likes to fight with my neighbor's bull, I have to keep the fences in good order.

Our farm is a family enterprise. We take good care of our land and cattle. It's a privilege to walk out to the pasture and just be able to watch them and their little calves interact. That's a reward all its own.

However, the bottom line is: I can afford to farm. I work to support my farming habit, which is one way to look at it. There is even a joke about this: Question: What does a farmer do after winning the million-dollar lottery? Answer: Farm until it's all gone.

This state of affairs is certainly not tenable in the long run. Often the word *sustainable* is used synonymously with *environmentally sound*—because it is so obvious that agriculture will not be sustainable if it damages the natural environment; Mother Nature can't be fooled. But it should be no less obvious that agriculture cannot be sustainable if it is not profitable.

For years now, as the numbers of farmers have decreased, I and many others have questioned just how sustainable the economics of industrial agriculture are. Industrial agriculture emphasizes maximum production. In contrast, sustainable agriculture emphasizes optimum production. Optimum production insists that the long term be factored in; it means that food production systems should balance society's food needs, profitability for the producer, natural resource conservation, and environmental protection.

It is often hard to successfully hit that balance.

However, if one looks at Kiowa County as an example of a typical rural county in the South, one would be hard-pressed to call maximum production a success. In Kiowa County, even fewer farmers claim agriculture as their principal occupation than in Le Flore County (see Figure 9.1). This is a sad fact in a county where agriculture has always been at the heart of life, and where no city is nearby to provide jobs.

The two counties present two faces of the profitability problem in agriculture. In Kiowa County, where the agriculture is centered on wheat and row crops, the problems have been similar to those all over farm country—problems of low commodity prices and pressures to get big or get out. The traditional road to profit of conventional industrial agriculture was to farm more land, get bigger machines, specialize in one crop, and use whatever chemicals it took to make a crop succeed.

On a macro level—the view from the plane flying overhead from New York to Los Angeles—this approach seemed to work, at least as far as production figures showed. Yields increased as fertilizer use increased. Farms got bigger and production boomed. On a micro level—the view from the front porch—the view was somewhat different. Some farmers hung in there, got bigger, and survived; many others lost. Along with the push for maximum production came its attendant

FIGURE 9.1. Kiowa County, Oklahoma, Census Count of Farms

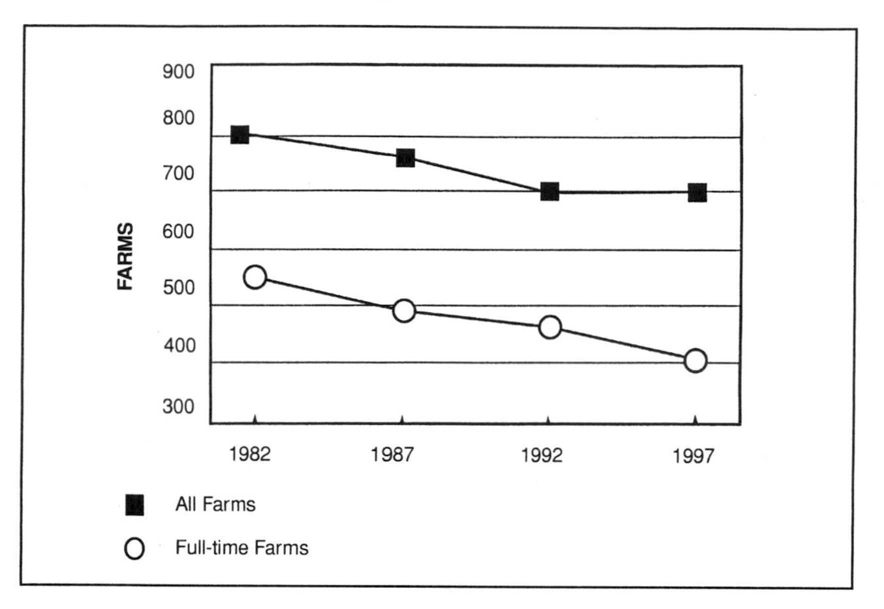

Source: Adapted from "1997 Census of Agriculture County Profile," United States Department of Agriculture, Oklahoma Statistics Service.

ills: soil degradation, water quality problems, production surpluses, low prices, and a decline in the rural economy.

The situation in Le Flore County looks different on the surface. But it is just as hard to make it in farming in this wetter, greener, hillier part of Oklahoma as it is in the harsher conditions further west. Row cropping here has been out since the 1950s; agriculture instead is centered on animals. There are no vast flat prairies here on which to grow cotton and wheat—just little flat patches with thin soil here and there among rocky woods.

The forty acres in front of my home that I use for a hay meadow illustrate the farming history of the area. The acreage was originally cleared during the 1930s by a man named Louie Scrivener. After sawing the timber, he grubbed the roots out—digging with a pick and a shovel around the roots tree by tree and then hooking a horse to the heavy roots and pulling them out of the ground. It's hard to imagine how long it took him to clear forty acres. And it says a lot about how tough people had it in farming in those days. He farmed with mules and grew corn.

By the time Senator Kerr came to the county, grubbing was out and 'dozing was in. Land clearing was done on a large scale with a bulldozer. Kerr began his love affair with the Black Angus and became the county cattle baron. He wasn't alone in his desire to raise fat cattle for market. By the early 1970s, cattle prices had skyrocketed, and there was a frenzied push by many to get in on the action. I had just gotten out of college, and I advised ranchers what I had been taught in school and what all the agricultural economists were saying: because of the way the feedlots were now operating, the days of boom and bust in cattle prices were over—what we knew as the cattle cycle would be nonexistent. Because of year-round feeding in the high plains of Texas and western Oklahoma, made possible by an aquifer (the Ogallala) that could grow irrigated grains to supply the feedlots, cattle would be produced every month of the year, and this would do away with seasonal or cyclical price fluctuations. Things had changed, the experts proclaimed, and prices were going to remain high.

As a result, the action centered on cattle. Timber was worthless, interest rates were low, and government loans were easy to obtain. The obvious thing to do was borrow money, clear timber, plant grass, buy cows, and cash in on the go-go cattle market.

I remember advising a large cattle rancher to buy cattle when they were selling at close to seventy-five cents per pound. Six or seven months later, the prices dropped by half. The market went bust. Suddenly, people had large debts they had to pay back on half the income—selling cattle for half of what they had paid for them. The supposedly defunct cattle cycle was still in force (after a few years of low prices, prices again recovered). This experience, while painful, taught me not to trust all of the economic reports coming out of the universities, and that I had better use my own common sense and be more realistic.

The ecological bust came a little later. During the boom, farmers borrowed money to clear land for their high-priced cattle. This clearing was done not just on flat bottomland but on the hillsides too. Chemicals were sprayed to defoliate the trees; where it was too rocky or steep to drill the grass seed, airplanes would fly over and drop fescue seed onto the now open land. The result: dead oak trees with vivid green grass below. The grass grew well for a few years, fed by the remaining fertile layer of dead leaves. After that, the fescue that likes deep, moist, bottomland best faded and thinned on the rocky soil. Our agriculture division specialists were sometimes called in to help at this point, when erosion had begun and gullies were forming. Their advice to the chastened rancher: replant trees.

Would-be farmers in southeast Oklahoma are now advised to go into contract chicken production. This is the new "place to be" in agriculture, a way to stay on the farm and have some kind of income from agriculture. The negatives of contract production—the big debt the farmer has to shoulder, the fact that he does not own the chickens and therefore has no equity in them, the uncertainty of contracts and amount of profit, the lack of decision-making power, negotiating power, and "fringe" benefits (health insurance, etc.)—all add up to a deal that has many risks.

Increase profitability and decrease risk is the final step in my eight steps to a healthy, enduring agriculture. One can increase profitability either by not spending as much to produce a crop or by receiving more for the crop at market. Decreasing risk means not borrowing as much money and being a good steward of the farm's resources so that they will be available over the long term—it means moving toward optimum production. It often means diversifying a farm's enterprises.

A PENNY SAVED

While American agriculture has generally followed the industrial paradigm in this century, with a few dissenters such as organic farmers in the Northeast and California looking to natural systems for inspiration, there has been another group of farmers in America who have not even been part of the debate, but have steadfastly gone their own way. They are the Amish.

The fact is that the Amish operate in the same price structure as other farmers, but are able to consistently make a profit from their small-scale farming enterprises. How do they do it? One answer: don't spend money. These farmers have done very well financially by growing food in old-fashioned ways, shunning chemical fertilizers and pesticides, avoiding or minimizing the use of mechanized equipment, planting windbreaks and plowing along the contour of hillsides to reduce erosion, and planting soil-regenerating cover plants in rotations with food crops.[6]

David Kline, a Holmes County, Ohio, Amish farmer, and his family milk about thirty Guernsey cows, an average-sized hand-milked dairy for the area. From April to November, while the cows traditionally produce $30,000 worth of milk, Kline spends about $1,000. The farm is a model of using on-farm resources—all of the grain and hay used for cows and horses are grown on the farm, and barn manure is used to fertilize the corn fields. In lieu of commercial fertilizer, clovers and alfalfa keep the pasture's energy level high. The Amish traditionally spend no money on diesel to run tractors, but use horses instead. As for the huge cost of new equipment, Kline joked in one article that "most of our equipment is so old it has to be carbon-dated to get an accurate age."[7]

The difference in profitability between an Amish farm and a conventional farm can be astounding and might be rather painful for a conventional farmer to contemplate. In the mid-1980s, the two were compared by Ohio State University experts, using detailed budgets. With the price of corn reckoned at $2.40 per bushel, a non-Amish farmer would have grossed $360 per acre against $393 in operating expenses for a net loss of $33 per acre. In contrast, the Amish would realize a net profit of $315 per acre.[8]

Although Amish agriculture appears to many to be caught in a time warp (albeit a profitable time warp), the average cost of running a conventional farm in the United States increased over 300 percent between 1950 and 1978 and continues to climb. What accounts for this? Inflation, higher prices for farm inputs, and, mostly, the increased use of fertilizers, pesticides, fuel, and machinery.

Taking a cue from the Amish, one of the base tenets of sustainable agriculture has been to reduce the use of off-farm inputs. By doing this, farmers can begin to reverse the trend of the past century where profit has been shifted from farmers or the production sector of agriculture to the input sector—suppliers of fertilizer, chemicals, and other goods and services to farmers. According to a study by University of Maine agricultural economist Stewart Smith, from 1910 to 1990, the farm sector's piece of the profitability pie has shrunk from 41 percent to 9 percent. At the same time, the input sector's share increased, from 15 to 24 percent, as did the marketing sector's share, from 44 to 67 percent.[9]

Strategies That Work

Farmers can reduce their use of chemicals and save money (and begin to be less controlled by high-priced technology) in a number of ways. Researchers in Arkansas have found that herbicides for soybean crops could be applied at a fraction of the recommended rates on the label with no loss in weed control or crop yield. Herbicide costs were reduced by 25 to 50 percent. Some ridge-till farmers are banding fertilizer in the ridge, and finding they can use about half of the broadcast rates of application, while maintaining yields and fertility. In New Jersey, corn growers successfully cut nitrogen use by an average of twelve dollars per acre by using a soil nitrogen test before applying the fertilizer. All told, the thirty growers in the trials saved about $15,000 in fertilizer costs. These cost savings are also good for the environment, decreasing the likelihood that nitrates or herbicide residues will end up in water supplies.

These approaches—emphasizing conservative use of inputs—are good first steps. The next step would be to redesign farms so that inputs are largely not needed. This was the idea behind the establishment of the USDA's Low-Input Sustainable Agriculture (LISA) program, now the Sustainable Agriculture Research and Education (SARE) program. Many LISA/SARE research projects have focused on reducing off-farm inputs by implementing alternative farming practices. Projects have focused on using crop rotations rather than pesticides to control pests; use of animal manures instead of chemical fertilizers; use of legumes in cover crops and rotation crops to provide nitrogen instead of chemical fertilizers.

Making these practices profitable to implement is key to their gaining wide adoption, and the SARE/LISA projects are designed to test for costs and savings. One such project demonstrated that farmers could save money on fertilizer and herbicide by using cover crops, and that further savings could be had by using cover crops that readily reseed themselves, thereby saving the expense of replanting them each year. (Cover crops are generally planted in the winter between crops to stop soil erosion. They are then killed in the spring. The cash crop is planted among the residue.)

The project was coordinated by Seth Dabney, of the National Sedimentation Lab in Oxford, Mississippi. Participants included thirteen researchers and six farmers across the South. The project looked to using cover crops in tandem with conservation tillage. Researchers identified two types of clover and found that farmers could save twenty-five dollars per acre seeding costs plus fifteen dollars per acre for fertilizer for three to four years after the initial planting. (If these seeds had carried the "terminator" gene, this project and these savings would have been impossible.)

Experiments with other cover crops showed that herbicide costs can be reduced because the cover crops shade out weeds when alive, and mulch them out after they have been killed. Researchers found that farmers can get the soil and water conservation benefits of no-till using cover crops and pay for the cover crop planting with savings in herbicide expense. This type of project points the way toward making conservation tillage practices such as no-till more ecologically and economically sound in the long term because herbicide use is reduced.[10]

Other research projects have demonstrated the big savings farmers can enjoy when they reduce inputs. A LISA project coordinated by Dr. Michael Smith at Oklahoma State University demonstrated that a cover crop of legumes under pecan trees in commercial orchards could reduce insecticide application by two-thirds, eliminate nitrogen needs, and reduce mowing costs. The result: a savings of twenty to forty dollars per acre, with no decreases in yield.

The list of profit-enhancing sustainable practices goes on and on. Some large corporations are already convinced. Multibillion dollar company Gallo Wine has converted more than half of its 10,000 acres of wine grapes to strictly organic methods. After a few years of transition when costs were higher, Gallo is reporting the same yields as ever, while spending less money per acre.[11]

Gallo says once the ecology of their vineyards was in balance, costs went down. Indeed, it is widely reported by organic and other sustainably oriented producers that production is more steady and stable than production from conventional farms and is better under environmental stress.

Whether growing row crop commodities such as cotton, tree crops such as pecans, horticultural crops such as grapes, or livestock, sustainable agriculture can help a farmer make more money. Rotational grazing is an example of a management strategy for cattle that replaces management for inputs and results in greater profit. Rotational or cell grazing also can increase revenue because more cattle can be raised on a given piece of land, while simultaneously improving its ecological health. Sound impossible? The conventional wisdom is that the only way to improve the health of rangeland is to take cows off it. But a switch in management may be more effective.

At a pasture management clinic at the Kerr Center in 1994, a large ranch in southern Oklahoma that had been operated in a conventional manner for thirty years was described. The cows had grazed continuously—that is, they had the free range of the place, eating what they pleased and when they pleased, and in doing so, overgrazed the ranch.

Implementing a rotational grazing system changed that. Pastures were divided up into fenced paddocks and cows were moved among them. With rotational grazing, the cattle are allowed to graze for a time in each paddock while the forage is watched carefully to make sure it is not grazed too short; then the paddock has a long rest period for the plants to recover.

The result: The cows were forced to eat "low seral" species of plants, that is, comparatively unpalatable annual grasses and weeds that would have otherwise have to be controlled by mowing or herbicides. These species declined by almost half, from 60 percent to 32 percent, in four years. More palatable grasses increased from 5 to 15 percent, and the total amount or forage or food available to cows increased by a whopping 300 percent.

This allowed the rancher to increase his stocking rate. Not only did he make more money from the increase in stocking rate, he saved money by substituting management (setting up the paddocks, moving the cattle) for inputs (herbicides, fertilizers). Comparisons of profit between a ranch practicing rotational grazing and a similar operation under continuous grazing showed gross income from conventional continuous grazing was 314 percent less than rotational stocking. Production costs—labor, equipment, and production inputs such as pesticides—also went down.[12]

Rotational grazing mimics what had been a natural cycle on the prairie, the action of the buffalo on the native grass ranges: The herd would graze for a short period of time, and move on in its migration, allowing the forage to recover and grow again. Today, on most ranches, the natural cycle is gone. When cattle continuously graze a pasture, grasses are cropped again and again, and the leaves cannot make the starches and sugars needed for plant growth and, in particular, healthy root development. When roots are weakened, growth declines and plants are weakened and are more susceptible to drought. Taking a cue from the natural cycle can help ranchers reestablish good forage for their cows and higher profits.

A PENNY EARNED

If one side of the profit equation is to reduce the amount of money spent in producing a crop, then the other side is to receive more for the crop at market. If one grows a conventional crop and markets it in a conventional way, it is difficult to get a lot more for the crop than the going rate. How can farmers break out of this trap? Some are doing it by growing higher-value crops or adding value to crops (i.e., processing the crop in some way). Others are marketing the crop in a way that allows a greater profit, such as through direct sales to consumers by subscription as in Community-Supported Agriculture (CSA) or at farmers' markets. Some are trying a combination of approaches.

Often when a farmer thinks of a crop that has a higher value than a grain crop, he or she thinks of fruits, vegetables, or nuts. Can a farmer make a living on a very small farm growing such crops? The experience of Alex and Betsy Hitt of Peregrine Farm in Graham, North Carolina, says yes. The Hitts are both full time on their five-acre farm. Peregrine Farm is highly diversified—producing cut flowers, small fruits, and vegetables, almost 200 different varieties. The Hitts have maximized their profits by marketing their crops through a farmers' market as well as directly to two natural foods grocery stores and a few restaurants, rather than selling to a wholesaler.

Lee Henry's Rockin' L-H Asparagus Farm near Stidham, Oklahoma, about seventy miles from the Kerr Center, is another success story, and a good example of "adding value" to a crop. His forty-acre asparagus patch yields thirty gourmet food products. About half of each year's output is shipped directly to 4,500 customers, including many gift and gourmet food stores around the United States and Canada. Henry also sells his products in person at state fairs and food shows. Henry reports his profits selling at wholesale are about 30 percent, but at retail they are about 130 percent.[13]

Henry is also a good example of a new kind of farmer—the farmer-entrepreneur. Farmers who go into direct marketing discover that they need to know more than how to grow a crop. They often must know how to process food, know how to manage production workers, know the regulations governing food businesses, and be skillful at marketing and promotion. These are not traditional farming skills; however, more and more farmers seem to be willing to take the plunge.

Other signs of change: The number of farmers' markets, where farmers sell their produce directly to consumers, continues to rise exponentially—up 37 percent from 1994 to 1996, and the total number of 2,410 is considered very conservative. Community-Supported Agriculture (CSA), also known as subscription farming, is also growing in popularity. Customers pay a farmer or a number of farmers in advance to provide produce and other farm products throughout the growing season. Customers share the risks of production by receiving larger quantities in a good year and less when times get tough. There are an estimated 1,000 or more CSAs in the United States, an arrangement that was virtually unheard of a dozen years ago.[14]

Joel Salatin of Polyface Farms in Virginia is probably the best known of this new wave of farmer-entrepreneurs. An articulate advocate for ecological farming, entrepreneurship, and creative thinking, Salatin has written widely about his farming enterprises. As mentioned, Salatin claims his approach to raising pastured poultry can net a hardworking farm couple $25,000 in six months on twenty acres. He is an advocate of diversity, producing cattle, eggs, turkeys, and broilers on the same pasture, and the profits to be gained from it—$5,000 per acre "stacking" the above crops on one acre. He direct markets to 400 customers, some of whom drive 400 miles to visit his farm.[15]

Although Salatin comes from a family of alternative farmers, others with more traditional farming backgrounds are also diversifying their enterprises with higher-value crops. In 1986, the Bradshaws of Innisfail, Alberta, Canada, began growing carrots, parsnips, and seed potatoes where once they produced only cereal grains, forages, or oil seeds. (They still grow those crops in rotation.) They mostly direct market their carrots through farmers' markets. Carrots that don't make the grade are fed to their Angus cow herd.[16]

The Salatin and Bradshaw farms illustrate two core principles for sustainable profitability. The first is diversification. Diversification has a number of advantages, both short term and long term. Indeed, when deciding what crops to grow, the farmer should look at both. The chief goal of diversification is stabilizing income—smoothing out highs and lows, which is particularly important for beginning farmers or farmers in precarious financial positions.

Reducing financial loss from fluctuating markets is a major benefit of diversification in the short term. One way to do this is to grow a diversity of crops that have different markets. Seldom do prices for all commodities rise and fall together—each has its own market and uses. Growing a variety of crops also helps protect farm income from weather or pest problems.

Diversify for Tomorrow

Today, farmers need to think about diversification a bit deeper than they are used to—to think about the long term, and what they can do to optimize, rather than just maximize, their production. Farmers with an eye toward future profitability should look to crops that require less fossil-fuel inputs, such as tree crops, as well as crops that can yield alternative fuels, such as soybeans. They should also look to build the fertility of their soil, because replacing natural fertility with chemical fertilizer will become less and less financially sustainable. Cover crops can hold soil and improve soil quality by adding organic matter and, if legumes are used, nitrogen. Adding animals to the mix on a farm can yield free manure to build fertility.

Every farm should also include an area for experimentation. Besides trying different tillage methods or experimenting with timing of seeding, the farmer might try potential new crops, to find out how adapted they are to that farm's conditions.

Farmers who fail to diversify may find themselves left behind. Indeed, diversity, as a key concept in sustainable agriculture, is likely to be part of the new mantra in the new century. The warning of the 1950s, "Get big or get out," may be replaced by the admonition "Diversify or die out." The time for new crops to come along and rejuvenate American agriculture seems to be at hand. Crops most people have never heard of, such as kenaf, milkweed, quinoa, and

(continued)

(continued)

meadowfoam, may be the commodities of the future. Indeed, several converging factors point to renewed interest in new crops, including concern over worldwide loss of biodiversity, low prices for major commodities, the sustainable agriculture and environmental movements, and consumer desire for new food and health products.[17]

As Jefferson once said, "The greatest service which can be rendered any country is to add a useful plant to its culture."

Standing Out from the Crowd

The other principle that seems to be emerging for small-scale farmers is emphasizing quality more than quantity in order to secure a market and perhaps get a premium price. The Bradshaws, for example, do not grow the variety of carrot, often grown in California, that is the industry standard. Instead, they grow a variety popular in Europe that they consider superior in sweetness and tenderness.

Organic crops are also considered by many consumers to be of higher quality, and growing organic crops can also yield a premium price. Everyone agrees that organics are a growth industry; statistics on the number of organic farmers, acreage in organic production, and organic sales to consumers may even underestimate the size of the trend. The USDA reports 4,060 certified organic producers in 1994, up from 2,841 in 1991. However, the Organic Farming Research Foundation estimates there are another 6,000 growers who meet general organic certification standards but do not certify their crops.[18] (Organic growers may be certified by a state agency or a private group. The process involves visits by the certifying agency as well as documentation by the grower. The USDA has recently issued national standards.) According to the latest figures from the *Organic Certifiers Directory,* the number of certified growers continues to grow and currently numbers over 6,000.[19]

The acreage in organic production is also increasing rapidly. In a 1994 survey, the USDA documented over a million acres in certified organic production, about double the 1991 acreage. (Again, by limiting the survey to officially certified farmers, the numbers may be drastically underestimated.)[20] Food crops, livestock feed, cotton, nursery/floral, tobacco, poultry, dairy cows, and beef cows are all being produced organically in the United States.

What Is Organic Agriculture?

Organic agriculture is popularly thought of as agriculture that uses no synthetic pesticides and fertilizers. However, there is more to it than that. An official definition (in part) from the National Organic Standards Board was devised in 1995:

> Organic agriculture is an ecological production management system that promotes and enhances biodiversity, biological cycles and soil biological activity. It is based on minimal use of off-farm inputs and on management practices that restore, maintain and enhance ecological harmony. . . .
>
> The principal guidelines for organic production are to use materials and practices that enhance the ecological balance of natural systems and that integrate the parts of the farming system into an ecological whole. . . . The primary goal of organic agriculture is to optimize the health and productivity of interdependent communities of soil life, plants, animals, and people.[21]

Another survey has found that although only 10 percent of Americans are regular purchasers of organic products, 60 percent indicated an interest in buying organics.[22] According to the Organic Trade Association, organic foods have posted annual sales growth of 20 percent during the 1990s, and the USDA projects a fourfold increase in organic food sales in the next decade. A multibillion-dollar market is also developing for organic goods overseas.[23]

The stats on who is farming organically provide a ray of hope for those of us who have been working to make farming on a small scale a viable choice. Organic farms are, on average, much smaller than their conventional counterparts, using about half the acreage of nonorganic operations. Certified organic vegetable growers tended to be quite small, the majority farming on less than five acres. Other encouraging signs: Most of these vegetable growers were between the ages of thirty-six and fifty-five, college educated, and, most important, listed farming as their primary occupation.[24]

Although there are farmers for whom the word "organic" is probably still akin to the word "communist," others are realizing that not only is the organic production of many crops possible, such production can be profitable. It may be that consumer demand for organic products is the key to getting farmers to change their farming systems away from the industrial model to a more sustainable model.

ECOLOGICAL ECONOMICS

The human species is part of nature. Its existence depends on its ability to draw sustenance from a finite natural world; its continuance depends on its ability to abstain from destroying the natural systems that regenerate this world.[25]

Although there are a number of things farmers can do production wise and marketing wise to cut costs, spread risks, and increase profitability, I believe that farmers and everyone else won't ever get a true sense of the costs of production until we account for certain "hidden" costs not included in the average farm budget.

These hidden costs affect farmers, society, and, perhaps most significant, generations to come. These hidden costs must be brought into the open and added up; otherwise, I believe we won't deal with them. It's simply too easy to ignore the fact that we are destroying our children's inheritance until it is too late. The price of commodities sold does not begin to cover the losses we incur in environmental damage and resource degradation in growing these commodities. This lack of whole-cost accounting subsidizes the price of food and keeps it cheap. Eventually natural resources will be used up and there will be a day of reckoning.

It is clear that a new, more rigorous cost/benefit analysis needs to be applied to agriculture in order to get a good idea of what we are gaining and what we are losing. In general, this is not being done. One example: comparing costs for various-sized hog "production units." Feeds, equipment, buildings, price of land, labor, management, and the cost of the animals themselves were included in the comparisons. The analysis found the large producer had a significant economic advantage over small and midsized units.[26]

Although this budget was probably accurate as far as it goes, it did not include any hidden costs, such as the costs associated with the polluted water that the waste from such farms can cause. On farms such as the 1,200-sow high-tech operation analyzed, the waste generated can be substantial (ten pounds of wet waste per day per hog), in this case 12,000 pounds per day. If a lagoon (where such waste is stored) breaks or overflows, the cost of fish and other aquatic and terrestrial life killed, cost for the extra processing that municipal water plants must do to make this water potable, and, last but not least, cost for the lost pleasure (and meals) of a fishing trip to the creek or river—none are generally factored in.

Then there are health costs from these hog units. Recently, the Minnesota Department of Health determined that toxic gas (hydrogen sulfide) emanating from a manure lagoon on one of the largest hog operations in that state posed a health hazard to nearby residents. High levels of the gas can reportedly cause respiratory problems, nausea, sore throats, and headaches.[27] None of these costs are accounted for.

Another example: the USDA has set the loss of five tons of soil per acre of topsoil as an acceptable rate. The average loss of American topsoil was about eight tons per acre in the late 1970s and early 1980s. A one-acre inch of topsoil weighs 160 tons. At the 1982 rate, an inch of soil from the average acre of cropland is lost every twenty years. Since experts estimate it takes anywhere from 300 to 1,000 years to produce an inch of topsoil,[28] this is a shocking loss. However, nowhere in current farm budgets is this loss of a valuable, hard-to-obtain resource counted as a cost of production—neither is the loss to the next generation in the decreased fertility of the land, nor the cost of the water pollution from silt.

As John Pesek points out in *Sustainable Agriculture Systems:*

> We [those of us in agriculture] have ignored the real cost of our technology at the farm level because we have not had to pay for the consequences (off-site effects), and society at large has not fully determined nor assessed costs of these effects on others and the environment. After all, the upland farmer does not directly have to pay for dredging silt from the Mississippi River nor does the farmer in north central Iowa have to worry about the nitrate in river water used for drinking in Des Moines.[29]

The costs of environmental degradation are borne by society, spread out, as former EPA head William Ruckelshaus has observed, and usually only apparent in the long term when the resource shows signs of serious strain or collapse. Then "the tragedy of the commons," as he calls it, is apparent, and the bill comes due.[30]

I have noticed that when farmers are asked about water pollution, they invariably answer that they have no desire to pollute—they, after all, have to drink the water, too. Despite this usually sincere disclaimer, the fact remains that farmers, at this point in time, generally do not pay the full cost of cleaning up any pollution they may have caused, however inadvertently.

These costs to the environment are as real as the cost of feed or seed, and someone pays, if not the farmer. As yet there has been little recognition of this problem by the government. This state of affairs may change, as some have taken on the arduous task of assigning value to these environmental costs. I don't envy those researchers their task; how do you decide what quality air is worth? What is an endangered species worth? What is good soil worth to future generations?

When I first read about these attempts to include environmental costs as costs of producing goods or services, I thought it was a pretty radical way of addressing resource problems. Back then, I still believed that technology, like some fairy godmother, would take care of any future problems. In my blind faith in American technology, I was a little like the agriculture scientists who believe that topsoil isn't needed at all—that the dense subsoil can be made to bloom given the proper application of fertilizer.

But not anymore. I have come to realize that natural resources are not free, and they are certainly not inexhaustible, as so many have so blindly supposed. I think of the head of the Bureau of Soils ninety years ago and his insistence that the soil of America was inexhaustible, while the hills of the South washed away all around him. The problem is just how to account for such costs.

These costs to the environment are as real as the cost of feed or seed, and someone pays, if not the farmer.

I have come to agree with those who propose that natural resources be depreciated just like handmade or manufactured assets, which are valued as productive capital and written off against their initial value as they depreciate. Under conventional accounting schemes, the use of natural resources, "free gifts from nature" with no investment required to create them, are not depreciated. In what has been called "national income accounting," the value of an asset is not its investment cost, but the present value of its income potential. Against this, the true value of depreciation should be charged. And what is the true value of depreciation? It is "the capitalized value of the decline in the future income stream because of an asset's decay or obsolescence."[31]

This is one way to make accounting sense of what is happening as our environment deteriorates. Farmers are accustomed to depreciating machinery, barns, and other capital investments, but not their soil, depleted of nutrients, organic matter, and microorganisms, which makes it less viable and of less value than it originally was when it was first farmed. Although machinery can be written off eventually and the cost of new tractors can be financed by new loans, the decline of basic resources is not so easily dealt with.

Assigning value to these "free" resources would go a long way toward changing the way we farm, and moving us toward sustainability. As William Vorley, an authority on the pesticide industry and sustainable agriculture, has written, if farmers were asked to pay for the total hidden costs of pesticide use including the pollution and the effects on human health, then perhaps pest management would shift toward nonchemical methods, currently perceived as uneconomical or inefficient. IPM strategies would also broaden, taking into account costs beyond the immediate costs of chemical versus nonchemical approaches.[32]

As compelling as these attempts are at an expanded, more complete accounting, they usually do not take into account a cost I feel is just as real: the human cost of lost livelihoods when farms go bust, rural communities decay, and cities swell with thousands from the rural hinterland. The costs to social services alone are considerable: increased welfare, homelessness, psychological and job counseling, job training, corrections. As farm families have left the land in the last fifty years, society has borne these costs without question.

Expecting farmers alone to pay for environmental costs would likely increase stress on the farm. Somehow, we must find a formula so that others involved in agriculture, perhaps the government and consumers, pay their share of these hidden costs.

Americans currently pay less for food than any other industrialized country. Americans spend 10.7 percent of their disposable personal income on food, down from 13.8 percent in 1970. This is largely due to per capita income going up 48 percent (adjusted for inflation) between those years.[33] The price of food does not reflect the hidden costs in food production. Perhaps it is time it did—affording farmers the opportunity to use more environmentally friendly, but possibly more expensive (at least in the short term), production methods.

Assigning value to these "free" resources would go a long way toward changing the way we farm, and moving us toward sustainability.

At stake is "intergenerational equity," the well-being of the next generations who are depending on us to make the right choices today. All the talk about money and power in the end comes down to a moral choice. Is it morally right to pass on to future generations a degraded resource base that will not support their aspirations? Is it right to think solely of ourselves? Aren't our children and our children's children entitled to an agriculture that is environmentally benign? Don't they have a right to fertile soils and clean water, to a resource-rich environment instead of a wasteland? Aren't they entitled to a world that is biologically diverse? Aren't they entitled to inherit greater knowledge of how ecosystems work and an appreciation of the varied characteristics of plants and animals? Aren't they entitled to a world where energy is used wisely and alternatives to nonrenewable fuels are being phased in so there is the least shock to everyone when fossil fuels run out?

Future generations deserve all this and more—farms and an economy that can be sustained and will sustain us. They should have an opportunity to farm if they want to, to be an independent small farmer, a member of a group that has traditionally provided a steadying influence on society. They should be able to enjoy a quality life in a rural area.

We worry about handing our children a gigantic budget deficit or a bankrupt Social Security system. Shouldn't we be just as worried about handing them an agriculture that doesn't work—in which the costs, human and environmental, are too high, and the power will soon be cut off?

I believe a sustainable agriculture is sound economics. It is only in the short term that there are conflicts. While we are up on the tightrope, doing our balancing act between yesterday and tomorrow, it may seem like we will never make it to the other side. But we will. We just have to make the right moves now.

Checklist for Farmers:
How to Increase Profitability and Reduce Risk

1. Diversify crops and livestock.
2. Substitute management for off-farm inputs.
3. Maximize the use of on-farm resources.
4. Keep machinery, equipment, and building costs down.
5. Add value to crops and livestock.
6. Try direct marketing (subscription farming, farmers' markets, farm stores, mail order).
7. Grow crops/livestock that receive premium prices (example: certified organics).

Chapter 10

Two Roads

I shall be telling this with a sigh
Somewhere ages and ages hence:
Two roads diverged in a wood, and I—
I took the one less traveled by,
And that has made all the difference.

Robert Frost, "The Road Not Taken"[1]

This book began with a courtroom scene and serious charges:

Industrial agriculture—defined as the current predominant system of agricultural production and its supporting establishment—stands accused in a three-part indictment, to wit:

of endangering the essential natural resources of soil, water, and life, thereby jeopardizing the future productivity of agriculture and the inheritance of our children;

of hooking farmers on fossil fuels, and the fertilizer and pesticides made from them, while downplaying the consequences of overusing such products;

of desolating rural America by bankrupting farmers and ignoring the well-being of rural communities, thus leaving them open to exploitation.

These crimes show a reckless disregard for the life and health of farmers, rural communities, and the natural world, jeopardizing our ability to feed an ever-growing population. As a result, the food security of our nation, and our world, is threatened.

Have I proved to your satisfaction the case against industrial agriculture? If not, let me take one more crack at it. My hometown, Cold

Springs, in Kiowa County, used to be a thriving rural community. There was a train depot, cotton gin, stores of all kinds, granite quarries, even a hotel. Because it was a beautiful area of mountains and creeks, folks from nearby cities would take the train to our town on weekends, drawn by the cold water from the springs and the well in the town. (Actually, the cold water was a bit of a hoax. On Friday nights, town boosters would slide two 300-pound blocks of ice into the well. After a customer remarked one Saturday that the water was the "coldest damn spring water" she had ever tasted, the town was dubbed Cold Springs.[2]

Farm people did their shopping in small towns such as Cold Springs. At one time, people did their shopping at small general stores such as the one our family frequented, owned by Mr. A. T. Henderson.

Henderson was a white-bearded gentleman, well versed in everything, it seemed. On the front porch of his store was a Coke machine, where one could get a pop for a nickel. Henderson had established his general store in 1913. At one time it was a prosperous business in a two-story building, selling everything from piece goods to binder repair. By the mid-1960s, though, Cold Springs had declined to the point that he was down to selling Coke and bread to the few families left in the area.

As a kid, I spent a lot of time sitting on the propane tank also out front, listening to my dad and Mr. Henderson talk about everything. I have fond memories of the old gent interrupting the preacher during Sunday service at the Friends (Quaker) meetings we attended. He was ready at any moment to blurt out a thought, a more appropriate Bible verse, or a brief testimony. We all loved it when he broke the silence commonplace at Quaker services. It was most delightful when a new or visiting preacher would take the podium, only to be pelted with Mr. Henderson's thoughts for the day.

Community stores like Mr. Henderson's, once common in rural America, are long gone, along with the rural communities and schools that once dotted the countryside. (So many of these are gone in Oklahoma that there is even a popular guidebook titled *Ghost Towns of Oklahoma*). Gone with them are the neighborliness and the sense of community that helped make life worthwhile. The unmaking of these towns was preceded by the unmaking of the family farms that supported them. In Kiowa County, agriculture has been declining as long as I can remember. And the decline continues: The number of farms was down almost 16 percent between 1982 and 1992. The county's population is aging, residents are increasingly dependent on government payments to live on, and many choose to leave, leading to an overall decline in population. As a result, even stores in the remaining "bigger" small towns are suffering. These days, substantial "retail leakages" occur, meaning that residents in Kiowa County drive out of the county to the nearest cities to shop and for entertainment.[3]

This is what happens when the family farm dies. A federal study of the most industrialized agricultural counties in California, Arizona, Texas, and Florida found that as farm size increases beyond that which can be worked and managed by a family, the quality of community life begins to deteriorate. Increasing concentration resulted in increased poverty, substandard living and working conditions, and a breakdown of social linkages between the rural communities that provided labor and the farm operators. The most extreme poverty was found in those counties with the most concentrated and productive agriculture.[4]

For the purposes of this book, I am going to draw a distinction between family farms and corporate agriculture. Although a family farm eludes precise definition, several characteristics determine whether a given place is a family farm or a branch of corporate agriculture. A family farm is owner operated, and entrepreneurial in decision making. Historically, it is family centered, using mostly family labor and management skills with help from the local labor pool. In general, a family farm is more resource conserving because of the linkage between the family and its land, and farming is viewed as a way of life (rather than just a way to maximize profits). The farmer has a commitment to certain values such as stewardship and a desire to pass on to the next generation a healthy, viable piece of land.

Unfortunately, too often these values have conflicted with conventional economics and have lost out. If our nation's agriculture was family farm oriented, rather than corporate agriculture oriented, our rural landscape would look quite different. Farms would be more uniform in size, more diversified, and markets would be open, allowing for healthy, equal competition. It is my firm belief that on a family farm—one that is not operating on the financial edge all the time, as is currently the case—stewardship of the land is much better.

The most extreme poverty was found in those counties with the most concentrated and productive agriculture.

Corporate agriculture is quite different. If owned by a nonfamily corporation, the workers and owners, both often from other places, have no real attachment to the land or the community. It is industrial agriculture taken to its logical end. It is large scale and concentrated, rather than being diversified. Production is centered on monoculture or a specific kind of livestock. Perhaps most important, it is not a way of life but strictly a business—entirely profit motivated, capital intensive, thriving in controlled markets where it has sufficient volume to take advantage of volume purchasing, and able to sell outside normal markets. Production processes are highly standardized, generally resource consumptive, and exploitive. Such businesses can exploit land or a community and move on, taking profits with them and buying into another community to do it again or transfer the money to a nonfarm business. Its sole purpose is to maximize shareholder values. How the land is treated and what will happen to it in the future are not primary concerns; short-term profits, however, are the primary concern.

Prime examples of corporate agriculture are the huge hog operations such as the ones owned by Seaboard Corporation in Texas County in the Oklahoma panhandle. This corporation moved into the panhandle after being enticed by myriad local, state, and federal financial incentives and set up "production units" and a slaughter plant that can handle four million hogs per year.[5] The county seat, Guymon, actively courted Seaboard, believing that this type of corporate industrial agriculture would revitalize the community and the county.

The results, though, have been decidedly mixed. Bitter social divisions have arisen between rural residents and hog boosters. Those who live near such facilities complain bitterly of the sickening odor that has drastically impaired their quality of life. Possible contamination of the water supply by the mountains of waste generated by these facilities is an ongoing concern. New jobs have been generated, but they are low paying, and have been largely filled by immigrants, not by local labor. Schools, social services, and law enforcement have been hard-pressed to cope with the arrival of these new residents.[6]

This is a prime example of desperation economics. The increasing domination of agriculture by large corporate interests is accelerating the destruction of healthy rural communities, causing them to take desperate steps to attract any and all jobs.

Corporate farms primarily provide low-income jobs. One report found that three jobs are lost for every one created by a new factory farm.[7] Although the family farmer returns nearly 90 percent of a year's income to the local area in taxes and farm-related purchases, corporate agriculture returns less than half, mostly in wages.[8] Corporate farms often access farm inputs from outside their communities and sell products to more distant markets as well. Corporate profits go elsewhere, too, often out of the state or even out of the country.

The worth of Mr. Henderson's thriving store, and stores like it, transcend nostalgia. They are an indicator of rural health. A study by Walter Goldschmidt of two farming communities in California's Central Valley tells the tale:

> One [farming community] was dominated by large farms, and the other was a community of small family farms. Where the family farm prevailed, Goldschmidt found a higher standard of living, better community facilities, such as streets and sidewalks, more parks, more stores with more retail trade, and twice the number of organizations for civic improvement and social recreation. . . . The small farm community had two newspapers where the other had only one. In short, the small farm community was a better place to live, perhaps because the small farm offered an opportunity for "attachment" to local culture and care for the surrounding land.[9]

ATTACHMENT TO THE LOCAL COMMUNITY AND CARE FOR THE SURROUNDING LAND

These are values that are essential to a healthy, enduring agriculture, a sustainable agriculture. Aren't they values also essential to a sustainable America and a sustainable world?

While the family farmer returns nearly 90 percent of a year's income to the local area in taxes and farm-related purchases, corporate agriculture returns less than half, mostly in wages.

If you agree, things we can do together as farmers and nonfarmers alike can transform our current system of agriculture into a system that is healthy and enduring. There are two roads in agriculture today. Which one we travel down will make all the difference. The road described in the indictment is the road to ruin—a dead end. Unfortunately, for the past fifty years, it has been the status quo, the road more traveled. Fortunately, an alternative way exists. It is a road that has been less traveled, but it is a road that goes somewhere.

I took my first step down the road less traveled in 1985 when the agricultural division of the Kerr Foundation became the Kerr Center for Sustainable Agriculture. After nearly fifteen years as a farm consultant, I had become convinced that agriculture as it was conventionally practiced in Oklahoma and around the nation could not be sustained.

Since then, I have become even more convinced that the hidden costs of industrial agriculture: the costs to soil, water, and life, to the natural environment, and farmers and rural communities, are not being adequately addressed by the industrial agriculture mainstream, never mind being tallied. If they are ever added up, I am convinced, industrial agriculture, with its monster machines and impressive yields, would not be nearly so impressive. And our food would not be nearly as cheap as it is.

Bucking the status quo has made my organization unpopular at times. In telling the flip side of the industrial agriculture success story, we rob some deans of agriculture of their material for speeches and charts.

So one has to be philosophical and take the long view. I feel like I'm carrying on in the pioneer spirit of Oklahoma—breaking new ground, looking for a better life. What keeps me going is the knowledge that the good earth will sustain us if we treat her right. (And, conversely, she will fail us if we fail her.) It has been a privilege to be part of the search for a sustainable agriculture, a search that has snowballed, and is growing in strength and credibility each year.

I feel like I'm carrying on in the pioneer spirit of Oklahoma—breaking new ground, looking for a better life. What keeps me going is the knowledge that the good earth will sustain us if we treat her right.

Down this alternative road is the next green revolution. The next green revolution will be different from the green revolution of the 1960s. It will stress optimum production instead of maximum production. Optimum production respects both short- and long-term needs. It integrates the immediate food and fiber needs of society and the immediate needs of farm families with the long-term necessity of maintaining natural resources and environmental health for everyone. The next green revolution will be holistic and will restore health to our farms, farm communities, and to an urban America that has become estranged from the natural world and the ancient cycles of sowing and reaping, which are essential for spiritual health.

The next green revolution will actually be a trio of revolutions: agricultural, ecological, and social/economic, all entwined. The agricultural revolution will look to enhance the productivity of our croplands not through fossil fuel-based inputs, but through stewardship: of soils, water, and life. Agriculture will be brought more into sync with natural cycles. Agriculture will loosen its ties to natural gas and oil and strengthen its ties to the sun, with systems that are powered by photosynthesis—infinitely renewable solar energy now—rather than by finite fossil fuels, energy stored from the past.

This agricultural revolution is also an ecological revolution, with the concept of an ecosystem broadened to include man-made systems such as farms—agroecosystems. Sustainable agroecosystems are modeled after natural ecosystems, with the goal to create as many natural balances in the system as possible. In addition to being powered by the sun through photosynthesis, agroecostems generate much or all of their fertility and pest resistance through complementary interactions among plants, animals, and soil organisms, and contain a wide variety of plant and animal species adapted to local conditions.

What will enable the first two revolutions to fully succeed is the third, the social/economic revolution. The real change that agriculture needs is, at its heart, social change. How society views food, farmers, and natural resources must change.

Fairness—whether farmers are getting a fair, adequate price for their crops, whether markets are free, whether the terms of contract production are fair, and whether government programs are equitable to all regardless of size of operation or race of operator—has not been a big concern of the industrial agriculture establishment or for society at large. The concentration of agricultural production into fewer and fewer hands has been largely accepted as inevitable. The control of the seed supply and of other key aspects of the agriculture system by fewer and larger corporations has also been accepted. In a sustainable agriculture, these trends must be reversed.

We also need to ensure that farmers and ranchers, who are arguably doing the most important work in the world in producing our food, are afforded a good quality of life: paid adequately, not expected to take on unreasonable financial risks or health risks from using toxic chemicals and working with dangerous machinery. It is a national shame how with frightening regularity come the farm crises, with their attendant tales of broken dreams and families, and farmer suicides.

How society views food is another paradigm shift. Consumers must be more knowledgeable about how their food is produced, whether the animals they eat are treated humanely, and whether the way the food is grown causes environmental pollution or displacement of farmers from the land. I believe that many consumers want to eat food that is produced in the most environmentally benign way possible. Beyond that, consumers are hungry for a more personal connection to farmers and farms—the explosion in popularity of farmers' markets and subscription farming demonstrates this. They want good, nutritious, locally produced food.

The next green revolution will actually be a trio of revolutions: agricultural, ecological, and social/economic, all entwined.

The road to a different kind of agriculture is a long, difficult road that will require many changes. I do not expect a sudden, drastic change in the way agriculture is done. I myself did not change my views with one blinding insight; I had many changes of heart.

The search for a healthy, sustainable agriculture is the greatest challenge of the new century. If we really are going to make the next green revolution, we must begin now; in agriculture, change comes slowly. A case in point is the recent (in the past fifteen years) switch from small square bales of hay to big round bales by cattle producers. The new round bale has advantages that make it desirable. However, buying a new baler to make the new kind of bale is expensive. Tractors to carry the round bales had to be bigger and fitted with special spears to lift them. In Oklahoma, it took about ten years for the change from square to round to take place—an average amount of time, say experts, for such changes in farming to be adopted.

The amount of time is compounded when the change is more radical and far reaching. Unfortunately, the clock is ticking while many in agriculture continue with business at usual. Some of those who received their training during the glory days of agriculture thirty and forty years ago continue to be hostile to the ideas of sustainable agriculture. Others have acknowledged the need for change. Even Robert E. Wagner, president emeritus of the Potash and Phosphate Institute (the organization which once promulgated the idea that organic farmers wanted to take us back to "a Tarzan life among the apes") has acknowledged that U.S. agriculture is "at a critical point." While criticizing proponents of low-input agriculture for their zeal, he nonetheless has called for a "middle ground" between all-organic systems and all-chemical systems. I agree that bridges must be built between the two camps so that progress can continue.[10]

One thousand years ago, the people threshing their wheat in the shadows of the manor houses and castles of Europe could not have foreseen the changes that would come in agriculture such as the combines that could crawl across fields like great wheat-eating dragons. Nor could the farmers hoeing their corn under the great ceremonial mounds of the Mississippi River and its tributaries have known that their crop one day would be grown in vast fields around the world, while they and their culture would be lost.

Well, maybe not completely lost. *Oklahoma* means, in the Choctaw language, "land of the red people." In Oklahoma, Euro-American pioneers mixed with Native Americans resettled here from all around the country. For decades, it appeared that the Native American cultures in the state would simply be overwhelmed by the power of the dominant American culture. But of late, Native Americans have reasserted themselves artistically, politically, and culturally, and Oklahomans of all backgrounds are being influenced. The time seems right for the traditional Native American ideas of respect for and connection to the earth, ideas that are implicit in sustainable agriculture, to be embraced by all of agriculture. Combined with the best ideas from the Euro-American farming tradition—the German farmers of the Shenandoah Valley during Jefferson's time, the Amish, the soil conservationists, and the biological/organic/regenerative farmers—we have an interesting mix of ideas to work with that may ultimately yield the next green revolution and a mainstream agriculture that is healthy and enduring.

And despite the sometimes-depressing trends in agriculture, I feel more optimistic about the possibility of change now than I have felt previously. Perhaps this is because of the progress I have seen made by people involved in sustainable agriculture. We are making a difference. We are still a small movement, but growing and evolving rapidly. We haven't gone away; we won't go away. We are influencing the debate on agriculture, and influencing what farmers do in the field every day.

The eight steps, or variations on them, are being taken to heart by more and more folks embarking on the journey down the road less traveled. (For the eight steps and a nutshell comparison of the approaches of industrial agriculture and sustainable agriculture, see Figure 10.1.) These farmers are those raising and selling free-range poultry and hoop-house hogs, forming cooperatives, running community-supported farms, and selling to farm markets. There are organic farmers and those who, although not organic, are trying to follow the eight steps as well as they can. There are farmers and researchers involved with the SARE program.

Although still only minimally funded (between twelve and fifteen million dollars annually), SARE has had an impact far beyond its modest budget. To date, the program has funded around 1,200 grants. The grants are competitive and administered through four regional offices. They fall into three categories: One type of grant is for research and education projects, usually led by interdisciplinary, multi-institutional, multistate research teams that also include farmers as participants. The second type of grant goes to producers to fund projects designed and

FIGURE 10.1. How They Compare

STEPS TO A HEALTHY, SUSTAINABLE AGRICULTURE	INDUSTRIAL AGRICULTURE	SUSTAINABLE AGRICULTURE
① ***Create and conserve healthy soil***	• Soil quality declining—soil erosion a chronic problem, organic matter not replenished, microbial activity damaged by farm chemicals, soil compacted by farming practices	• Soil quality a central concern—soil protected from erosion by cover crops, residue, low-impact tillage, and conservation measures such as windbreaks; organic matter continually added, farming methods and smaller sized machinery keep soil loose and friable
	• Conventional tillage, conservation tillage combined with heavy chemical use	• Conservation tillage techniques combined with biofriendly management to cut use of chemicals
② ***Conserve water and protect its quality***	• Water is mined from dropping aquafers, agricultural chemicals degrade water supplies and threaten aquatic life	• Farming methods conserve water and soil moisture and protect surface and groundwater from pollutants and sediment
	• Conservation structures and areas take a back seat to more production	• Conservation is a top priority; terraces, buffer strips, riparian buffers and other conservation structures, practices, and areas incorporated into the farm
③ ***Manage organic wastes to avoid pollution***	• CAFOs concentrate large amounts of animal wastes in one place, overloading the ability of the area to utilize it and also increasing chances of spills and water pollution	• Animal wastes provide nutrients for growing crops without polluting watersheds; smaller numbers of animals are raised on integrated farms where they are part of a diversified system

	INDUSTRIAL AGRICULTURE	SUSTAINABLE AGRICULTURE
④ ***Select plants and animals adapted to the environment***	• With large amounts of inputs, farmers can raise nonadapted crops	• Farmers raise animals and plants adapted to the existing environment
⑤ ***Encourage biodiversity***	• Genetic engineering further narrows genetic diveristy	• Time-honored, traditional breeding programs look to preserve genetic diversity
	• Monoculture is the norm: farms are plowed fence row to fence row, wild "unused" areas are put into production, only the most productive few crop varieties or livestock breeds are raised	• Diversity is the norm: of habitats, livestock, crops, wild plant and animal species, and of genetics within crop and livestock species
⑥ ***Manage pests with minimal environmental impact***	• Therapeutic approach—chemicals are used routinely to control pests	• The use of toxic chemicals for pest control is minimized and ecologically based, benign management and cultural practices used
⑦ ***Conserve nonrenewable resources***	• Powered by finite fossil fuels: fertility and pest control needs filled by agricultural chemicals	• Powered by the sun: fertility and pest control largely provided by cycling of plants and animals in the system using rotations, cover crops, trap crops, resistant crops
	• Use of fossil fuels encouraged	• Renewable energy resources (biofuels, solar) substituted when possible and conservation of fossil fuels encouraged
	• Food production is centralized in a few regions that specialize in certain crops, which are shipped around the nation and world	• Food production is decentralized to encourage local, biodiverse, environmentally adapted food systems which save fossil fuels

	INDUSTRIAL AGRICULTURE	SUSTAINABLE AGRICULTURE
⑧ ***Increase profitability and reduce risk***	• Small and medium-sized farms are marginalized, pressure is on farmers to increase the size of their operations	• Small and medium-sized family farms generate equitable returns so that farmers can protect natural resources, stay in business over the long term, and have a good quality of life
	• Decreasing numbers of farmers are needed	• Believes that we need more farmers on the land
	• Large corporations control farmers and markets through contracts and vertical integration	• Free markets prevail and farmers have control over how they farm
	• The farm is viewed solely as an agribusiness	• The farm is viewed holistically, with the quality of life of the farm family one part of a whole
	• Short-term profit is the focus	• Long-term consequences of farming methods are given equal weight to short-term profit
	• Farmers have little control over profits	• Through adding value, direct marketing, and other profit-enhancing approaches, farmers can set prices and assert more control over the prices they receive
	• Dependent on high-priced technology	• Works with low-input biological processes
	• Large scale, so debt and risk high	• Smaller-scale enterprises do not demand as much borrowed money

implemented by them. And there are professional development projects designed to train agricultural information providers, such as extension personnel, in sustainable agriculture techniques and concepts. The purpose of these grants is to increase knowledge about, and help farmers and ranchers adopt, sustainable practices that are profitable, environmentally sound, and beneficial to society.

In most cases, farmers and ranchers collaborate with researchers to create proposals for SARE funding. SARE has changed the lives of many of these farmers. A case in point is Tom Trantham. The Pelzer, South Carolina, dairy farmer discovered SARE at a time when his farming practices needed an overhaul. "I had focused so tightly on production I didn't see the rest of the world," he recalls. "I would wake up some mornings and hope the place had burned down."[11]

In 1996, his dairy was in its third season as a SARE grazing project. The project is, according to Trantham, "proving what I knew in my heart at the beginning—sustainable systems work. Because that system works, I have more time. Time to enjoy my cows instead of worrying about staying ahead of their feed bills. Time to spend with my family."[12]

We began this book with the words of Thomas Jefferson, that contradictory figure, who said that the small landholders are the most precious part of the state. Jefferson was a man who participated in the plantation system, where cotton and tobacco were grown until the soil was depleted, yet he loved passionately the lush fertility of his Monticello garden and orchards. He was a plantation man, a slaveholder, yet he believed fervently in the worth of the small, independent farmer. Jefferson is much like Americans today, with our divided allegiances. On the one hand, we are taken with quantity: with industrial agriculture, with monoculture, and maximum production no matter what the human or environmental costs. On the other hand, we crave quality: a "greener" agriculture, one that favors diversity, equity, and a healthy environment.

Let's hope that we too will be remembered, as Jefferson is, for our more ethical, farsighted actions. Change *can* happen. The next green revolution is not out of our reach; in fact, it's already begun. We are beginning a new drama in a new century. There is reason for hope: "The earth," as Jefferson wrote in 1813, "belongs to the living, not to the dead."[13]

Notes

Chapter 1

1. Department of Agriculture, Natural Resources Conservation Service, *A Geography of Hope,* Program Aid 1548, 1996, p. 36.
2. Ibid., p. 39.
3. Ibid., p. 11.
4. Ibid., p. 33.
5. "SWCS Policy Position Statement: Sustainable Agriculture," *Journal of Soil and Water Conservation 50*(6), (November/December 1995), p. 635.
6. William P. Cunningham and Barbara Woodworth Saigo, *Environmental Science—A Global Concern* (Dubuque, IA: William C Brown, 1990), p. 428.
7. Department of Agriculture, *A Geography of Hope,* p. 40.
8. American Rivers, (n.d.) *America's Twenty Most Endangered Rivers of 1998,* Available online: <http://www.amrivers.org> [April 1999].
9. Brian Ford, "Lake Eucha Quality Talk Grim," *Tulsa World 92*(342), (August 21, 1997).
10. "Fertilizers Growing a Dead Zone in Gulf," *Tulsa World* 94(142), (January 25, 1999).
11. Department of Agriculture, *A Geography of Hope,* p. 44.
12. "Endosulfan Responsible for Alabama Fish Kill" (Alabama Department of Environmental Management, August 24, 1995, press release), from Pesticide Action Network North America Updates Service, February 23, 1996, quoted in Charles M. Benbrook, Edward Groth III, Jean M. Halloran, Michael K. Hansen, and Sandra Marquardt, *Pest Management at the Crossroads* (Yonkers, NY: Consumers Union, 1996), p. 64.
13. Environmental News Network, News Archive (August 30, 1995), *Study Finds High Nitrate Levels in Agricultural Areas.* Available online: <http://www.enn.com/enn-news-archive/1995/08/083095/08309519.asp> [February 17, 2000].
14. R. Douglas Hurt, *American Agriculture: A Brief History* (Ames, IA: Iowa State University Press, 1994), pp. 340-341.
15. U.S. Water News Online (April 1996), *High Plains Drought Endangers Agriculture, Ogallala Aquifer,* Available online: <http://www.uswaternews.com/archive/96/supply/ogallala.htm> [January 1999].
16. Geoffrey C. Saign, *Green Essentials* (San Francisco: Mercury House, 1994), p. 111.
17. Hope Shand, *Human Nature: Agricultural Biodiversity and Farm-Based Food Security* (Ottawa: Rural Advancement Foundation International, 1997), p. 21.

18. Donald E. Bixby, Carolyn J. Christman, Cynthia J. Ehrman, and D. Phillip Sponenberg, *Taking Stock: The North American Livestock Census* (Blacksburg, VA: McDonald and Woodward Publishing, 1994), p. 19.

19. Ibid., p. 13.

20. Ibid., p. 14.

21. Ibid., pp. 19-20.

22. Ibid., p. 20.

23. Gary Gardner, "Shrinking Fields: Cropland Loss in a World of Eight Billion," *Worldwatch Paper* 131, July 1996, p. 6.

24. Ibid., p. 5.

25. Department of Agriculture, Economic Research Service, Natural Resources and Environment Division, *1995 Nutrient Use and Practices on Major Field Crops,* AREI Updates No. 2, May 1996.

26. Ibid.

27. Arsen J. Darnay, Ed., *Statistical Record of the Environment* (Detroit: Gale Research, 1992).

28. Charles M. Benbrook, Edward Groth III, Jean M. Halloran, Michael K. Hansen, and Sandra Marquardt, *Pest Management at the Crossroads* (Yonkers, NY: Consumers Union, 1996).

29. Amory B. Lovins, L. Hunter Lovins, and Marty Bender, "Energy and Agriculture," in *Meeting the Expectations of the Land,* Eds. Wes Jackson, Wendell Berry, and Bruce Colman (San Francisco: North Point Press, 1984), p. 71.

30. Ibid., p. 68.

31. "Annual Energy Outlook 1999—Market Trends—Oil and Gas Prices," Department of Energy, Energy Information Center, pp. 3-7. Available online: <http://www.eia.doe.gove.oiaf/aeo99/gas.html> [January 2000].

32. John Gever, Robert Kaufmann, David Skole, and Charles Vorosmarty, *Beyond Oil* (Cambridge, MA: Ballinger Publishing, 1986), p. 27.

33. Benbrook et al., *Pest Management at the Crossroads,* p. 2.

34. For the 20,000 figure, see World Health Organization/UNEP, *The Public Health Impact of Pesticides Used in Agriculture* (Geneva: 1990), quoted in Lester R. Brown, Christopher Flavin, and Hal Kane, *Vital Signs 1996* (New York: W.W. Norton, 1996), p. 108.

35. Saign, *Green Essentials,* pp. 307, 308.

36. Department of Agriculture, Economic Research Service, Natural Resources and Environment Division, *Agricultural Research,* AREI Updates No. 5 Revised, 1995.

37. Ibid.

38. Robert A. Hoppe, Robert Green, David Banker, Judith Z. Kalbacher, and Susan E. Bently, *Structural and Financial Characteristics of U.S. Farms, 1993,* Department of Agriculture, Economic Research Service, Agriculture Information Bulletin, no. 728, p. 30.

39. Hurt, *American Agriculture,* p. 317.

40. Department of Agriculture, *Structural and Financial Characteristics,* p. iii.

41. Department of Agriculture, Economic Research Service, Agriculture and Rural Economy Division, *Economic Indicators of the Farm Sector: National Financial Summary 1992,* ECFIS 12-1, calculated from Table 49, quoted in Patricia

Allen, *The Human Face of Sustainable Agriculture,* University of California, Santa Cruz, Sustainability in the Balance Series, Issue Paper No. 4, 1994, p. 4.

42. Neil D. Hamilton, "Agriculture without Farmers? Is Industrialization Restructuring American Food Production and Threatening the Future of Agriculture?" *Northern Illinois University Law Review 14*(3), (summer 1994), p. 630.

43. John E. Ikerd, "Sustainable Agriculture: Farming in Harmony with the Biosphere," in *Sustainable Agriculture: Enhancing the Environmental Quality of the Tennessee Valley Region through Alternative Farming Practices,* University of Tennessee Extension Service, EC 1022-750-9/92, quoted in Janet Bachman, "Sustainable Agriculture and Rural and Urban Communities: What Are the Connections?" *KCSA Newsletter 18*(11), (November 1992), p. 2.

44. David Strawn, "Rural Development Conference," (paper presented at the Kerr Center Rural Development Conference, Shawnee, OK, June 26, 1997), pp. 9-10, 25-26.

45. Hurt, *American Agriculture,* p. 390.

46. Department of Agriculture, *Structural and Financial Characteristics,* p. iii.

47. D. MacCannell, "Agribusiness and the Small Community," background paper to *Technology, Public Policy and the Changing Structure of American Agriculture,* Office of Technology Assessment (Washington, DC: Government Printing Office, 1983) quoted in Janet Bachman, "Sustainable Agriculture and Rural and Urban Communities: What Are the Connections?" *KCSA Newsletter 18*(11), (November 1992), p. 3.

48. Rick Welsh, *Reorganizing U.S. Agriculture: The Rise of Industrial Agriculture and Direct Marketing* (Greenbelt, MD: Henry A. Wallace Institute for Alternative Agriculture, 1997), pp. 9, 10.

49. Department of Rural Sociology, "Concentration of Agricultural Markets" (University of Missouri: October 1997, unpublished photocopy).

50. "Biotechnology," *Doane's Agricultural Report 60*(37-1), (September 12, 1997), p. 1.

51. Gigi DiGiacomo and Harry Smith, "A Look at Who's Gobbling Up America's Food Dollar: Agribusiness Shortchanging Farmers and Consumers," *Farm Aid News 3*(12), (June 28, 1995), p. 1.

52. T. Hall, "In Cuisine of the 90's, the Farmer Is the Star," *The New York Times,* July 8, 1992, B4, quoted in Janet Bachman, "Sustainable Agriculture and Rural and Urban Communities: What Are the Connections?" *KCSA Newsletter 18*(11).

53. Liberty Hyde Bailey, *The Holy Earth* (Lebanon, PA: Sowers Printing, 1915).

Chapter 2

1. John Pesek, "Historical Perspective," in *Sustainable Agriculture Systems,* Eds. J. L. Hatfield and D. L. Karlen (Boca Raton, FL: Lewis Publishers, 1994), p. 2.

2. United States Department of Agriculture Study Team on Organic Farming, *Report and Recommendations on Organic Farming,* July 1980, p. 78.

3. House Appropriations Subcommittee on Agriculture, Rural Development, and Related Agencies, Testimony of Jess Andrews III, *Low Input Agricultural Re-*

search, Hearings on Subsection C of Title 14 of the 1985 Farm Bill, Washington, DC, April 8, 1987.

4. House Appropriations Subcommittee on Agriculture, Rural Development, and Related Agencies, Testimony of Brian Chabot, *Low Input Agricultural Research, Hearings on Subsection C of Title 14 of the 1985 Farm Bill,* Washington, DC, April 8, 1987.

5. Ibid.

6. House Appropriations Subcommittee on Agriculture, Rural Development, and Related Agencies, Testimony of Charles A. Francis, *Low Input Agricultural Research, Hearings on Subsection C of Title 14 of the 1985 Farm Bill,* Washington, DC, April 8, 1987.

7. Bruce D. Smith, *The Emergence of Agriculture* (New York: Scientific American Library, 1995), p. 13.

8. Stephen E. Ambrose, *Undaunted Courage: Meriwether Lewis, Thomas Jefferson, and the Opening of the American West* (New York: Simon and Schuster, 1997) pp. 32, 33.

9. Ibid., p. 33.

10. Richard R. Harwood, "A History of Sustainable Agriculture," in *Sustainable Agricultural Systems,* Eds. Clive A. Edwards, Rattan Lal, Patrick Madden, Robert H. Miller, and Gar House (Delray Beach, FL: St. Lucie Press, 1990), p. 5.

11. Ibid., p. 9.

12. Ibid., p. 8.

13. Sir Albert Howard, *An Agricultural Testament* (New York: Oxford University, 1943), p. B.

14. Edward H. Faulkner, *Plowman's Folly* (New York: Grosset and Dunlap, 1943).

15. Harwood, "History," in *Sustainable Agricultural Systems,* p. 6.

16. Ibid., p. 10.

17. *Facts from Our Environment* (Atlanta: Potash Institute of North America, 1972).

18. Study Team, *Report,* p. xi.

19. William D. Ruckelshaus, "Toward a Sustainable World," in *The Energy-Environment Connection,* Ed. Jack M. Hollander (Washington, DC: Island Press, 1992), p. 368.

20. Jack M. Hollander, introduction to *Energy-Environment.*

21. Ward Sinclair, "How's That Again, Mr. Deputy Secretary?" *The Washington Post*, April 29, 1988, p. A-5.

22. Ibid.

23. Neill Schaller, "Federal Policies to Fully Support Sustainable Agriculture Research and Education," in *Agricultural Conservation Alternatives: The Greening of the Farm Bill.* Ed. A. Ann Sorensen (DeKalb, IL: American Farmland Trust Center for Agriculture in the Environment, 1994), p. 73.

24. Ibid., p. 77.

25. U.S. General Accounting Office, *Sustainable Agriculture: Program Management, Accomplishments and Opportunities,* GAO/RCEDS-92-233, 1992, p. 3.

26. J.F. Parr, R.I. Papendick, I.G. Youngberg, and R.E. Meyer, "Sustainable Agriculture in the United States," in *Sustainable Agricultural Systems,* Eds. Clive A.

Edwards, Rattan Lal, Patrick Madden, Robert H. Miller, and Gar House (Delray Beach, FL: St. Lucie Press, 1990), p. 51.

27. National Research Council, *Report on Alternative Agriculture,* (Washington, DC: National Academy Press, 1989), p. 2.

28. University of California Sustainable Agriculture Research and Education Program, (n.d.) *What Is Sustainable Agriculture.* Available online: <http://www.sarep.ucdavis.edu/sarep/concept.html#Natural Resources>. [November 6, 1997].

29. Vivian M. Jennings and John Ikerd, "National Perspective on Agricultural Sustainability" (paper presented at the American Society of Agronomy North Central Branch Meeting in conjunction with the Registry for Environmental and Agricultural Professional (REAP) Workshop, Des Moines, IA, August 1, 1994), p. 1.

30. Ibid.

31. Cokie Roberts and Steven V. Roberts, "Hunger: A Startling Crisis," *USA Weekend,* March 27-29, 1998, p. 5.

32. Susanna B. Hecht, "The Evolution of Agroecological Thought," in Miguel A. Altieri, *Agroecology: The Scientific Basis of Alternative Agriculture* (Boulder, CO: Westview Press, 1987) pp. 17-18.

33. Chuck Hassebrook, "Is There a Place for Biotechnology?" (paper presented at Future Farms conference, Oklahoma City, OK, February 8, 2000).

34. Charles M. Benbrook, remarks to Sustainable Agriculture Scoping Group, President's Council on Sustainable Development, Washington, DC, April 28, 1994.

35. Charles Francis and Garth Youngberg, "Sustainable Agriculture—an Overview," in *Sustainable Agriculture in Temperate Zones,* Eds. Charles A. Francis, Cornelia Butler Flora, and Larry D. King (New York: Wiley, 1990).

36. Harwood, "History," in *Sustainable Agricultural Systems,* p. 4.

37. Geoffrey C. Saign, *Green Essentials* (San Francisco: Mercury House, 1994), p. 137.

38. Pesek, "Historical," in *Sustainable Agriculture,* p. 14.

39. University of California, *What is Sustainable,* p. 2.

40. Dennis T. Avery, *Saving the Planet with Pesticides and Plastics: The Environmental Triumph of High-Yield Farming* (Indianapolis, IN: Hudson Institute, 1995).

41. GAO, *Sustainable Agriculture: Program,* p. 25.

Chapter 3

1. Wendell Berry, "Essay," *Audubon 95*(2), (March/April 1993), p. 104.

2. F. Dwain Phillips, "Potential for Wind Erosion Is at 20 year high," (U.S. Department of Agriculture, National Resources Conservation Service, Stillwater, OK, March 12, 1996, press release).

3. Donald Worster, *Nature's Economy: A History of Ecological Ideas* (Cambridge, UK: Cambridge University Press, 1977), pp. 225-228.

4. Wheeler McMillen, *Feeding Multitudes* (Danville, IL: Interstate Printers and Publishers, 1981), p. 35.

5. William P. Cunningham and Barbara Woodworth Saigo, *Environmental Science—A Global Concern* (Dubuque, IA: William C Brown, 1990), p. 226.

6. McMillen, *Feeding Multitudes,* p. 33.

7. Theodore Roosevelt, December 8, 1908 quoted in Harold Evans, "Saving America the Beautiful," in Random House, *The American Century, Old and New Americans 1889-1910,* p. 2. Available onlinez: <http://www.randomhouse.com/features/americancentury/oldnew.html> [December 2000].

8. Ibid., pp. 33-34.

9. Ibid., pp. 34-35.

10. Department of Agriculture, *Natural Resources Service (April 1995), Natural Resources Inventory: A Summary of Natural Resource Trends in the U. S. between 1982 and 1992,* Available online: <http://www.greatplains.org/resources/nrisumm/LOSESOIL.HTM> [March 1999].

11. William P. Cunningham and Barbara Woodworth Saigo, *Environmental Science, A Global Concern,* Third Edition (Dubuque, IA: William C. Brown, 1995), p. 227.

12. Department of Commerce, National Technical Information Service, Economic Research Service, *Soil Erosion and Conservation in the United States: An Overview,* AIB718 (October 1995), table 3, p. 5.

13. Department of Agriculture, Natural Resources Service (April 1995), *Natural Resources Inventory: A Summary of Natural Resource Trends in the U. S. between 1982 and 1992,* Available online: <http://www.greatplains.org/resource/nrisumm/SLIDE. HTM> [March 1999].

14. Shaun Schafer, "Erosion Program Enrollment Blowin' in the Wind," *Tulsa World 92*(14), (September 25, 1996), sec. E, p. 3.

15. Mike Nichols, "Agriculture Observes Earth Day Every Day," (Oklahoma Farm Bureau, Oklahoma City, Oklahoma, April 21, 1998, press release), p. 1.

16. Joseph A. Cocannouer, *Farming with Nature* (Norman, OK: University of Oklahoma Press, 1954), p. 15.

17. Marty Strange, *Family Farming: A New Economic Vision* (Lincoln, NE: University of Nebraska Press, 1988), p. 4.

18. L.M. Thompson and F.R. Troch, *Soil and Soil Fertility* (New York: Oxford University, 1978), p. 111.

19. Maria Filimonova, "Life Underground," *Kerr Center Newsletter* (November/December 1996), p. 9.

20. Sir Albert Howard, *An Agricultural Testament* (New York: Oxford University Press, 1943), p. 28.

21. Charles Darwin, Paul H. Barrett, and R.B. Freeman, *The Formation of Vegetable Mould Through the Action of Worms with Observations on Their Habits,* Vol. 28 (1881; reprint New York: New York University Press, 1990).

22. Mary F. Fauci and R.P. Dick, "Soil Microbial Dynamics: Short- and Long-Term Effects of Inorganic and Organic Nitrogen," *Soil Science Society of America Journal 58* (May-June 1994), pp. 801-806.

23. George Williams and Patricia Y. Williams "Nutrition Affects Susceptibility of Corn to Borer Damage,"*HORTIDEAS 15*(8), (August 1998), p. 89.

24. Howard, *Testament,* p. 30.

25. Cocannouer, *Farming with Nature,* p. 7.

26. Howard, *Testament,* p. 4.

27. Dan Crummett, "Ridge-Till Coming to Oklahoma," *Oklahoma Farmer-Stockman 105*(4), (April 1992), pp. 5-6.

28. "Newswatch," *Progressive Farmer 112*(12), (December 1997), p. 6.

29. Dan Crummett, "Can You Afford Not to No-Till?" *Oklahoma Farmer-Stockman 112*(4), (April 1999), p. 8.

30. "Farm Facts and Fancies," *The Furrow 102*(8), (November 1997), p. 40.

31. Don Comis, "Farmers Now Part of the Global Warming Solution as U.S. Agriculture Becomes Net Carbon Sink" (Department of Agriculture, Agricultural Research Service News Service, May 17, 1999, press release).

32. Center for Rural Affairs, "No-till and Sustainability," *Beginning Farmer Newsletter,* March 1998, p. 3.

33. Department of Agriculture, Sustainable Agriculture Research and Education Program, North Central Region, *1997 Annual Report* (Lincoln, NE, 1998).

34. Staff, wire reports, "State Resists Conservation Tillage," *Tulsa World, 93*(46), (October 28, 1997), sec. E, p. 3.

35. U.S. Department of Agriculture, Soil Conservation Service, *Farming with Residues* (Des Moines, IA, 1991).

36. Worster, *Nature's Economy,* p. 224.

37. Ibid., pp. 228, 229.

38. Le Flore County Conservation District, " Long Range Total Resource Conservation Program July 1994-June 1999" (Le Flore County Conservation District, Poteau, OK, report photocopy 1999), p. 14.

39. Department of Agriculture, Natural Resources Service, (April 1995), *Natural Resources Inventory: A Summary of Natural Resource Trends in the U. S. between 1982 and 1992,* Available online: <http://www.greatplains.org/resource/nrisumm/NONFED.HTM> [March 1999].

40. Department of Agriculture, Natural Resources Conservation Service, *A Geography of Hope,* Program Aid 1548, 1996, p. 33.

41. Ibid., p. 33.

42. Roscoe R. Snapp and A.L. Newmann, *Beef Cattle,* Fifth Edition (New York: John Wiley and Sons, Inc., 1966), p. 41.

43. Potash supplies: IMC Global, Inc., (n.d.) *World Crop Nutrients and Salt Situation,* Available online: <http://www.imcglobal.com/cropsalt/nak20cap.htm>. Phosphorus supplies: IMC Global, Inc., (n.d.) *World Crop Nutrients and Salt Situation,* Available online: <http://www.imcglobal.com/cropsalt/k20rgncap.htm>.

44. Cunningham and Saigo, *Environmental Science,* p. 227.

45. Amory B. Lovins, L. Hunter Lovins, and Marty Bender, "Energy and Agriculture," in *Meeting the Expectations of the Land,* Eds. Wes Jackson, Wendell Berry, and Bruce Colman (San Francisco: North Point Press, 1984).

Chapter 4

1. John McPhee, "Encounters with the Archdruid," 1971 quoted in Barbara K. Rodes and Rice Odell, *A Dictionary of Environmental Quotations* (New York: Simon and Schuster, 1992), pp. 187-188.

2. Department of Commerce, National Technical Information Service, Economic Research Service, *Soil Erosion and Conservation in the U.S.: An Overview,* AIB718 (October 1995), table 6, p. 18.

3. Janine Castro and Frank Reckendorf, "Effects of Sediment on the Aquatic Environment: Potential NRCS Actions to Improve Aquatic Habitats," working paper No. 6, Natural Resources Conservation Service, Oregon State University, Department of Geosciences, August 1995.

4. U.S. Environmental Protection Agency, *National Water Quality Survey,* (1994) quoted in U.S. Department of Agriculture, Natural Resources Conservation Service, *Water Quality,* RCA Issue Brief 9, March 1996.

5. Ibid.

6. U.S. Environmental Protection Agency, *National Water Quality Inventory,* report to Congress (Washington, DC, 1990) quoted in Arsen Darnay, Ed., *Statistical Record of the Environment* (Detroit: Gale Research, Inc.), p. 17.

7. Geoffrey C. Saign, *Green Essentials* (San Francisco: Mercury House, 1994), pp. 135-136.

8. Environmental Working Group, (n.d.) *Pouring It On: Nitrogen Use and Sources of Nitrate Contamination,* Available online: <http://www.ewg.org/pub/home/reports/Nitrate/NitrateUse.html> [July, 1999].

9. Sir Albert Howard, *An Agricultural Testament* (New York: Oxford University Press, 1943), p. 4.

10. Department of Agriculture, Natural Resources Conservation Service, *A Geography of Hope,* Program Aid 1548, 1996, p. 44.

11. Robert F. Kennedy Jr. "I Don't Like Green Eggs and Ham!" *Newsweek* (April 26, 1999), p. 12.

12. National Campaign for Sustainable Agriculture, *Environmental Degradation and Public Health Threats from Factory Farm Pollution,* Fact Sheet No. 1 (November 17, 1998).

13. Julie Anderson, "States Monitor Hog Waste's Effects on Groundwater," *Omaha World Herald 134*(72), (December 28, 1998).

14. "Biting Perspectives," California Sustainable Agriculture Research and Education Program Newsletter: *Inquiry in Action,* 21, (January 1999).

15. William P. Cunningham and Barbara Woodworth Saigo, *Environmental Science, A Global Concern,* Third Edition (Dubuque, IA: William C Brown, 1995), p. 430.

16. Department of Agriculture, Natural Resources Conservation Service, *Water Quality,* RCA Issue Brief 9 (March 1996).

17. Ann Ziebarth, *Well Water, Nitrates and the 'Blue Baby' Syndrome Methemoglobinemia,* Available online: <http://ianrwww.unl.edu/IANR/PUBS/NEBFACTS/nf91-49.htm> [July 1999].

18. Anna M. Fan, Calvin C. Wilhite, and Steven A. Book, "Evaluation of the Nitrate Drinking Water Standard with Reference to Infant Methemoglobinemia and Potential Reproductive Toxicity," in *Regulatory Toxicology and Pharmacology,* Vol. 7 (1987), pp. 135-148 quoted in Ann Ziebarth, *Well Water, Nitrates and the 'Blue Baby' Syndrome Methemoglobinemia.* Available online: <http://ianrwww.unl.edu/IANR/PUBS/ NEBFACTS/nf91-49.htm> [July 1999].

19. Charles Oliver, "Paying U.S. Farmers to Pollute: Subsidies Spur Increased Pesticide, Fertilizer Use," *Investors Business Daily—National Issue,* October 6, 1995.

20. Department of Agriculture, *A Geography of Hope,* p. 48.

21. Environmental Protection Agency, "Pesticides in Ground Water: Background" document WH-550G (Washington, DC: GPO, 1986) quoted in Jim Bender, *Future Harvest* (Lincoln: University of Nebraska Press, 1994), p. 112.

22. Oliver, *Investors Business Daily.*

23. Department of Labor, Occupational Health and Safety Administration, "Agricultural Operations," p. 1, Available online: <http://www.osha.slc.gov/SLTC/agriculturaloperations/> [January 2, 2001].

24. Department of Agriculture, *A Geography of Hope,* p. 47.

25. Boyd Kidwell with Earl Manning and Karl Wolfshohl, "The Big Push for Buffer Strips," *Progressive Farmer 113*(12), (November 1998), pp. 20-21.

26. T.C. Jacobs and J.W. Gilliam, "Riparian Losses of Nitrate from Agriculture Drainage Waste," *Journal of Environmental Quality 14* (1985), pp. 472-478 quoted in Oklahoma Cooperative Extension Service, Oklahoma State University, and Oklahoma Consevation Commission, *Riparian Area Management Handbook* (Stillwater, OK: Oklahoma State University, 1998), p. 2.

27. Kidwell et al., *Progressive Farmer,* p. 21.

28. Jim Suszkiw, "Natural Environmental Protection Agents," *Agricultural Research 46*(2), (February 1998), p. 8.

29. Oklahoma Water Resources Board, *Synopsis of the Oklahoma Comprehensive Water Plan,* Pub. 94-S, January 1980, p. 7.

30. Leon New (n.d.), *Opportunities to Maximize the Utilization of Water by Irrigators,* P & R Surge Systems, Available online: <http://www.prsurge.com/otmtuowb.htm> [July 1999].

31. Ibid.

32. Ibid.

33. Monica Manton Norby, "Improved Cropping Systems to Protect Groundwater Quality," *The Aquifer 13*(2), (August 1998), p. 6.

34. Oklahoma Water Resources Board, *Synopsis of the Oklahoma Comprehensive Water Plan,* p. 4.

35. Jim Britton, Le Flore County Area Poultry Extension Specialist, telephone conversation with Maura McDermott, April 15, 1997.

36. Ibid.

37. Oklahoma Water Resources Board, Water Quality Programs Division, *Diagnostic and Feasibility Study of Wister Lake, Executive Summary,* (Oklahoma City: Oklahoma Water Resources Board, May 1, 1996), pp. 1-2.

38. Ibid, p. 1.

39. Ibid, p. 4.

40. Oklahoma Conservation Commission, *Poteau River Comprehensive Watershed Management Program,* Oklahoma FY-94, 319h work plan, March 1995, p. 21.

41. Ibid.

42. Oklahoma Water Resources Board, *Diagnostic and Feasibility Study of Wister Lake,* p. 2.

43. Department of Agriculture, *A Geography of Hope,* p. 48.

44. National Campaign for Sustainable Agriculture, *Environmental Degradation and Public Health Threats from Factory Farm Pollution.*

45. Howard, *Testament,* p. 4.

46. Oklahoma Conservation Commission, *Poteau River Comprehensive Watershed Management Program,* p. 22.

47. Joel Salatin, introduction to *Pastured Poultry Profits* (Swoope, VA: Polyface, Inc., 1993), p. vi.

48. Ibid., pp. 106-107.

49. Anne Fanatico, "New Free Range Poultry Guidebook Covers Production, Processing and Promotion," *ATTRA News* 5(3), (June 1997).

50. Mark and Nancy Moulton, (n.d.) "Deep Straw in Hoops: Managing Manure in Concert with the Natural, Social and Economic Environment," *Purdue University Manure Management,* Paper 37, Available online: <http://kyw.ctic.purdue.edu/FRM/ManureMgmt/Paper37.html> [July 1999].

51. National Campaign for Sustainable Agriculture, *Sustainable Alternatives to Factory Farm Animal Production,* Fact Sheet No. 2, November 1998.

52. Jim Cantrell, *A Low-Investment Swine System* (Poteau, OK: Kerr Foundation Agricultural Division, 1985).

Chapter 5

1. Kenneth Williams, "Five Reasons Why You Should Consider Sustainable Farming Practices," *KCSA Newsletter 17* (3), (March 1991), p. 3.

2. Susanna B. Hecht, "The Evolution of Agroecological Thought," in *Agroecology: The Scientific Basis of Alternative Agriculture* (Boulder, CO: Westview, 1987), p.6.

3. At the fastest rate of formation (one-half ton per year), this amount of topsoil would take ten years to form. United States Department of Agriculture, Natural Resources Conservation Service, Soil Quality Institute, *Effects of Soil Erosion on Soil Productivity and Soil Quality,* Soil Quality-Agronomy Technical Note No. 7 (August 1998), p. 1.

4. J. Patrick Madden, "Goals and Realities of Attaining a More Sustainable Agriculture" (paper presented at the conference, Environmentally Sound Agriculture, Orlando, FL, April 1994), p. 1.

5. Wheeler McMillen, *Feeding Multitudes* (Danville, IL: Interstate Printers and Publishers, 1981), p. 17.

6. Oklahoma Biodiversity Task Force, Norman L. Murray, Editor, *Oklahoma's Biodiversity Plan: A Shared Vision for Conserving Our Natural Heritage* (Oklahoma City: Oklahoma Department of Wildlife Conservation, 1996, draft copy), p. 21.

7. *Description of the bottomland forest in this region,* Ibid., p. 21.

8. Robert S. Kerr, *Land, Wood, and Water* (New York: Fleet Publishing Corporation, 1960), p. 359.

9. Will Lathrop, "The Kerr Center Ranch, 1987-1996," (Kerr Center for Sustainable Agriculture, Poteau, OK, photocopy), p. 1.

10. Chris Agee, "Fungus, Grazing and Stocker Options," Kerr Center for Sustainable Agriculture Newsletter 23 (3), p. 1.

11. Lathrop, "The Kerr Center Ranch," p. 9.

12. Victor Davis Hanson, *Fields Without Dreams* (New York: The Free Press, 1996), p. 98.

13. Helga Olkowski and William Olkowski, "Design of Small-Scale Food Production Systems and Pest Control," in *Sustainable Food Systems,* Ed. Dietrich Knorr (Westport, CT: AVI Publishing Company, 1983).

Chapter 6

1. Prince Bernhard of the Netherlands, speech to the Young Presidents Organization, May 6, 1974, quoted in Barbara K. Rodes and Rice Odell, *A Dictionary of Environmental Quotations* (New York: Simon and Schuster, 1992), p. 9.

2. Paul Sobocinski, "The Crisis That Monocropping Built," *Land Stewardship Letter 16* (5), (November 1998), p. 3.

3. Brian DeVore, "Biodiversity and Agriculture: A House Divided," *Land Stewardship Letter 16* (5), (November 1998), p. 9.

4. Ibid., p. 1.

5. Department of the Interior, Fish and Wildlife Service, *Why Save Endangered Species,* Brochure p. 2.

6. Geoffrey C. Saign, *Green Essentials* (San Francisco: Mercury House, 1994), p. 113.

7. John R. Anderson Jr., Peter T. Bromley, H. Michael Linker, and Duane F. Newman, "Effects of Sustainable and Conventional Agriculture on Farm Wildlife," in *1994 Sustainable Agriculture Research and Education (SARE)/Agriculture in Concert with the Environment (ACE) Annual Report* Ed. Gwen Roland (Griffin, GA: SARE, 1995), p. 93.

8. James Paul, "Internalizing Externalities: Granular Carbofuran Use on Rapeseed in Canada," *Ecological Economics 13*(3), (June 1995), pp. 181-184 quoted in Tracy Irwin Hewitt and Katherine R. Smith, "Intensive Agriculture and Environmental Quality: Examining the Newest Agricultural Myth," *Report from the Henry A. Wallace Institute for Alternative Agriculture* (Greenbelt, MD, September 1998), p. 4.

9. Council for Agricultural Science and Technology (CAST) (n.d.), *Benefits of Biodiversity*, Available online: <http://www.cast-scienc.org/biod/biod_ch.htm> [September 13, 1999], p. 8.

10. Department of the Interior, *Why Save Endangered Species.*

11. Hewitt and Smith, "Intensive Agriculture and Environmental Quality," p. 4.

12. R.E. Warner, "Illinois Farm Programs: Long Term Impacts on Terrestrial Ecosystems and Wildlife-Related Recreation, Tourism and Economic Development," Report prepared for the Research and Planning Division, Illinois Department of Energy and Natural Resources (Springfield, IL, 1991) quoted in Hewitt and Smith, "Intensive Agriculture and Environmental Quality," p. 4.

13. SARE/ACE *1994 Annual Report,* p. 93.

14. J. Kim Kaplan, "Conserving the World's Plants," *Agricultural Research* 6(9), (September 1998), p. 8.

15. CAST, *Benefits of Biodiversity,* p. 8.

16. Ibid., p. 4.

17. Lynne Trewe, "The Other Endangered Species," *Permaculture Drylands Journal* 28 (spring 1997), p. 18.

18. The American Livestock Breeds Conservancy 1999 Breeders Directory, (Pittsboro, NC: American Livestock Breeds Conservancy, 1999).

19. Saign, *Green Essentials,* p. 114.

20. Donald E. Bixby, American Livestock Breeds Conservancy membership letter to Maura McDermott, November 10, 1997.

21. Trewe, "The Other Endangered Species," p. 1.

22. Donald E. Bixby et al., *Taking Stock: The North American Livestock Census* (Blacksburg, VA: McDonald and Woodward Publishing Company, 1994), p. 40.

23. Ibid., p. 39.

24. CAST, *Benefits of Biodiversity,* p. 2.

25. Saign, *Green Essentials,* p. 114.

26. Hope Shand, *Human Nature: Agricultural Biodiversity and Farm-Based Food Security* (Ottawa: Rural Advancement Foundation International, 1997), p. 21.

27. Food and Agriculture Organization (FAO) of the United Nations, *State of the World's Genetic Resources for Food and Agriculture* (Rome, 1996), p. 22 quoted in Shand, *Human Nature,* p. 21.

28. Ibid.

29. Saign, *Green Essentials,* p. 114.

30. Shand, *Human Nature,* p. 21.

31. George Anthan, "Seed Bank System May Wither," *Muskogee Phoenix and Times Democrat,* April 19, 1998, p. C-10.

32. Shand, *Human Nature,* p. 37.

33. Laura Tangley, "How to Create a Life Without Sex," *U.S. News and World Report 127*(4), (July 26, 1999), p. 41.

34. Science and Technology News Network, (n.d.), *Killer Corn—Local Tips,* Available online: <http://www.stm2.net/pages11/killercorn/ltips.html> [September 7, 1999], p. 2.

35. Manjula V. Guru and James E. Horne, *Biotechnology: A Boon or a Curse* (Poteau, OK: Kerr Center, 2000), p. 2.

36. Genetic Resources Action International (GRAIN), (n.d.) *Intellectual Property Rights and Biodiversity: The Economic Myths,* Available online: <http://www.grain.org/publications/gtbc/issue3.htm> [September 1999] quoted in Guru and Horne, *Biotechnology,* p. 4.

37. Gyorgy Scrinis, (n.d.), Colonizing the Seed, *Canberra Organic Growers Society, Inc.* Available online: <http://www.netspeed.com.au/cogs/gen3.htm> [August 1999] quoted in Guru and Horne, *Biotechnology,* p. 4.

38. Robert Rhoades, "The Incredible Potato," *National Geographic 176*(5), (May 1982), quoted in Phil Williams, "A Man for All Seasons (and Time Zones)," *Georgia Alumni Record 73*(1), (December 1993).

Chapter 7

1. Rachel Carson, *Silent Spring* (Greenwich, CT: Fawcett, 1962), p. 24.

2. Pete Daniel, *Breaking the Land* (Chicago: University of Illinois Press, 1985), p. 6.

3. Ibid., p. 8.
4. Ibid., p. 300, n. 11.
5. Wheeler McMillen, *Feeding Multitudes* (Danville, IL: The Interstate Printers and Publishers, 1981), p. 76.
6. Ibid., p. 307.
7. A.L. Aspelin, *Pesticide Industry Sales and Usage: 1992 and 1993 Market Estimates,* U.S. Environmental Protection Agency, Washington, DC, June 1994 quoted in Toni Nelson, "Efforts to Control Pesticides Expand," in *Vital Signs* (New York: W.W. Norton, 1996), p. 108.
8. Marty Strange, Liz Krupicka, and Dan Looker, "The Hidden Health Effects of Pesticides," in *It's Not All Sunshine and Fresh Air: Chronic Health Effects of Modern Farming Practices* (Walthill, NE: Center for Rural Affairs, 1984), p. 55.
9. Department of Agriculture, Economic Research Service, *Agricultural Resources and Environmental Indicators,* Agricultural Handbook no. 705, quoted in Charles M. Benbrook, Edward Groth III, Jean M. Halloran, Michael K. Hansen, and Sandra Marquadt, *Pest Management at the Crossroads* (Yonkers, NY: Consumers Union, 1996), figure 2.2, p. 44.
10. Benbrook et al., *Pest Management at the Crossroads* (Yonkers, NY: Consumers Union, 1996), p. 42.
11. A.L. Aspelin, *Pesticide Industry Sales and Usage: 1994 and 1995 Market Estimates—Preliminary,* quoted in Benbrook et al., *Pest Management at the Crossroads,* p. 41.
12. Department of Agriculture, Economic Research Service, Natural Resources and Environment Division, *Pest Management on Major Field Crops,* AREI Updates No. 1, February 1997.
13. W. Joe Lewis, J.C. van Lenteren, Sharad C. Phatak, J.H. Tumlinson III, "A Total System Approach to Sustainable Pest Management," in *Proceedings of the National Academy of Sciences 94* (November 1997), p. 1223.
14. Aspelin, *Pesticides Industry Sales,* quoted in Benbrook et al., *Pest Management at the Crossroads,* p. 42.
15. Benbrook et al., *Pest Management at the Crossroads,* p. 42.
16. Ibid., p. 32.
17. Ibid., p. 2.
18. William P. Cunningham and Barbara Woodworth Saigo, *Environmental Science—A Global Concern* (Dubuque, IA: William C. Brown, 1990), Table 12.2, p. 246.
19. Aspelin, *Pesticides Industry Sales,* quoted in Benbrook et al., *Pest Management at the Crossroads,* p. 34.
20. "Doane's Focus Report: Planting Decision '97," *Doane's Agricultural Report 60*(8), (February 21, 1997), p. 5.
21. Harold Willis, *The Coming Revolution in Agriculture* (Wisconsin Dells, WI: self-published, 1985), p. 68.
22. "Kinder Killers: Many Little Hammers," *The American Gardener, (*March/April 1997), p. 50.
23. John Gever et al., *Beyond Oil* (Cambridge, MA: Ballinger Publishing, 1986), p. 170.
24. Cunningham and Saigo, *Environmental Science,* pp. 241-242.

25. Department of the Interior, U.S. Geological Survey Circular 1090, *Persistence of the DDT Pesticide in the Yakima River Basin Washington* (Washington, DC: GPO, 1993).

26. SimpleLife, (n.d.), *Organic Cotton Exhibit,* Available online: <http://www.simplelife.com/organiccotton/001-ORGANIC%20COT.html> [May 19, 1998].

27. Geoffrey C. Saign, *Green Essentials* (San Francisco: Mercury House, 1994), p. 306.

28. Pesticide Action Network of North America, (n.d.) *Exports of Hazardous Pesticides from U.S. Ports Increases,* Available online: <http://www.panna.org/panna/> [May 26, 1998].

29. Victor Davis Hanson, *Fields Without Dreams* (New York: The Free Press, 1996), p. 70.

30. Ibid.

31. For the 20,000 figure, see World Health Organization/UNEP, *The Public Health Impact of Pesticides Used in Agriculture* (Geneva: 1990); for 25 million figure see J. Jeyaratnam, "Acute Pesticide Poisoning: A Major Problem," *World Health Statistics Quarterly 43*(3), 1990, both quoted in Lester R. Brown et al., *Vital Signs 1996* (New York: W.W. Norton, 1996), p. 108.

32. Robert Repetto and Sanjay S. Baliga, *Pesticides and the Immune System: The Public Health Risks* (Washington, DC: World Resources Institute, 1996), p. 16.

33. Benbrook et al., *Pest Management at the Crossroads,* p. 59.

34. Strange, Krupicka, and Looker, "Hidden Health" in *Not All Sunshine,* pp. 57-59.

35. Benbrook et al., *Pest Management at the Crossroads,* p. 3.

36. Strange, Krupicka, and Looker, "Hidden Health" in *Not All Sunshine,* p. 58.

37. Sheila Hoar Zahm and Aaron Blair, "Pesticides and Non-Hodgkins Lymphoma," *Cancer Research 52* (October 1, 1992): 5485s-5488s, quoted in Tracy Irwin Hewitt and Katherine R. Smith, "Intensive Agriculture and Environmental Quality: Examining the Newest Agricultural Myth," *Report from the Henry A. Wallace Institute for Alternative Agriculture* (Greenbelt, Maryland, September 1998), p. 5.

38. Repetto and Baliga, *Pesticides and the Immune System,* p. 49.

39. Robert Repetto and Sanjay S. Baliga, *Executive Summary: Pesticides and the Immune System: The Public Health Risks* (Washington, DC: World Resources Institute, 1996), p. 2.

40. Benbrook et al., *Pest Management at the Crossroads,* p. 32.

41. Lewis et al., "A Total System Approach," in *Proceedings,* p. 12,243.

42. Allen Jobes, "Crop Rotation for Disease and Insect Control," in *Kerr Center Newsletter 23*(1), (1997), p. 2.

43. Doris Stanley, "Testing Diversified Orchard Ecosystems: Cover Crops, Mixed Trees Encourage Beneficial Insects," *Agriculture Research 44*(1), (January 1996), pp. 18-19.

44. Bruce D. Smith, *The Emergence of Agriculture* (New York: Scientific American Library, 1995), p. 13.

45. Lewis, et al., "A Total System Approach," in *Proceedings,* p. 12,243.

46. Helga Olkowski and William Olkowsi, "Design of Small-Scale Food Production Systems and Pest Control," in *Sustainable Food Systems,* Ed. Dietrich Knorr (Westport, CT: AVI Publishing Company, 1983), p. 144.

47. Jan Suszkiw, "Biological Warfare Against Beet Armyworms," *Agricultural Research 46*(1), (January 1998), p. 17.
48. McMillen, *Feeding Multitudes,* p. 146.
49. N. Waissman Assadian, L. Corral Esperanza, C. Ponce, "Organic Cotton Production—1993-1997," (Texas A&M Research Center, El Paso, Texas, 1999, photocopy), p. 29.
50. Ibid.
51. Lewis, et al., "A Total System Approach," in *Proceedings,* p. 12,246.
52. Texas Organic Marketing Cooperative "Mission Statement," Texas Organic Marketing Cooperative, O'Donnell, Texas, 1998, photocopy, p. 2.
53. Harold Willis, *The Coming Revolution in Agriculture* (Wisconsin Dells, WI: self-published, 1985), p. 83.
54. Donald L. Vogelsang, *Local Cooperatives in Integrated Pest Management,* U.S. Department of Agriculture Farmer Cooperative Service Research Report 37, February 1977, p. 1.
55. Ibid., p. 1
56. Dennis R. Kenney, "Biological Control: Its Role in Sustainable Agriculture," in *Leopold Letter 8*(4), (winter 1996), p. 8.
57. Ibid., p. 3.

Chapter 8

1. University of California Sustainable Agriculture Research and Education Program, (n.d.). *What is Sustainable Agriculture.* Available online: <http://www.sarep.ucdavis.edu/sarep/concept.html#Natural Resources> [November 6, 1997].
2. John P. Holdren, "Energy Agenda for the 1990's," *Energy Use Trends,* p. 380, table 13.1, in *The Energy-Environment Connection,* Ed. Jack M. Hollander (Washington, DC: Island Press, 1992).
3. Jeanne M. Devlin, "A Century of Oil," *Oklahoma Today 47*(3), (July-August 1997), pp. 62-63.
4. "Up and Down Wall Street—Crude Awakening," *Barron's,* April 7, 1996, quoted in "Smooth Sailing or Rough Seas Ahead for Energy Prices?" Royale Energy Report, (May, 1996), pp. 1-2. Available online: <http://www.royl.Com/nl/5-96/ html#GEOLOGY> [January, 1997].
5. Colin J. Campbell and Jean H. Laherrere, "The End of Cheap Oil," *Scientific American 278*(3), (March 1998), p. 83.
6. Holdren, "Energy Agenda," p. 382.
7. Campbell and Laherrere, "The End of Cheap Oil," pp. 78-83.
8. Dennis T. Avery, *Saving the Planet with Pesticides and Plastics: The Environmental Triumph of High Yield Farming* (Indianapolis, IN: Hudson Institute, 1995), p. 231.
9. Gregg Easterbrook, "Good Pollution," *The New Yorker* (October 20/27, 1997), p. 80.
10. Avery, *Saving the Planet,* p. 232.
11. Robert Frost, "Mowing," *The Poetry of Robert Frost,* Ed., Edward Connery Lathem (New York: Henry Holt, 1969), p. 25.

12. Michael Perelman, "Efficiency in Agriculture," in *Radical Agriculture,* Ed. Richard Merrill (New York: Harper and Row, 1976), pp. 72-73.

13. Ibid., p. 79.

14. Amory B. Lovins, L. Hunter Lovins, and Marty Bender, "Energy and Agriculture," in *Meeting the Expectations of the Land,* Eds. Wes Jackson, Wendell Berry, and Marty Bender (San Francisco: North Point Press, 1984), p. 70.

15. William P. Cunningham and Barbara Woodworth Saigo, *Environmental Science, A Global Concern,* Third Edition (Dubuque, IA: William C Brown, 1995), p. 230.

16. U.N. Food and Agricultural Organization, *Estimated Energy Intensiveness of Food Crops* (1987) quoted in Cunningham and Saigo, *Environmental Science,* p. 231, table 11.4.

17. John Gever, Robert Kaufmann, David Skole, and Charles Vorosmarty. *Beyond Oil* (Cambridge, MA: Ballinger Publishing, 1986), p. 152.

18. Ibid., pp. 159-160.

19. Cunningham and Saigo, *Environmental Science,* p. 230.

20. "Doane's Focus Report: Planting Decisions '97," *Doane's Agricultural Report 60*(8), (February 21, 1997), p. 5.

21. Oklahoma State University, Department of Agricultural Economics, "Bottomland Soybean Budget for Southeast Oklahoma," (March 15, 1994).

22. U.S. Bureau of the Census, "Farm Income and Expenses: 1980-1995," *Statistical Abstract of the United States 1997,* 117th Edition (Washington, DC, 1997), p. 671, table 1095.

23. Kenneth Williams, "Five Reasons Why You Should Consider Sustainable Farming Practices," *KCSA Newsletter17*(13), (March 1991), p. 2.

24 "Low-Input Farming Systems: Benefits and Barriers," *Seventy-Fourth Report by the Committee of Government Operations* (Washington, DC: GPO, 1986), p. 16.

25. Ibid., pp. 16-17.

26. Dan Crummett, "Ridge-Till Coming to Oklahoma," *Oklahoma Farmer-Stockman 105*(4), (April 1992), pp. 5-7.

27. Ibid., p. 6.

28. Ibid., pp. 5-7.

29. Kenneth Repogle, "Sustainable Family Farming," (Remarks to the President's Council on Sustainable Development, Coweta, Oklahoma, September 23, 1997).

30. Lovins et al., "Energy and Agriculture," p. 74.

31. Crummett, "Ridge-Till Coming to Oklahoma," p. 5.

32. Frederick Kirschenmann, "Switching to a Sustainable System" (Windsor, ND: Northern Plains Sustainable Agriculture Society, 1988).

33. S.B. Hill and Jacques Nault, "Towards a More Energy Efficient Cash Crop Farm," *Sustainable Farming 6*(1), (fall 1995), pp. 8-11.

34. Luke Elliott and Judi Fox, *Solar Water Pumping,* Meadowcreek Technical Brief No. 2 (Fox, AR: Meadowcreek, 1992).

Chapter 9

1. "Smashing the Hourglass," *Land Stewardship Letter 17*(1), (January-March 1999), p. 13.

2. William D. Ruckelshaus, "Toward a Sustainable World," in *The Energy-Environment Connection,* Ed. Jack M. Hollander (Washington, DC: Island Press, 1992), pp. 368-369.

3. Robert A. Hoppe, Robert Green, David Banker, Judith Z. Kalbacher, and Susan E. Bently, *Structural and Financial Characteristics of U.S. Farms, 1993,* Department of Agriculture Economic Research Service, Agriculture Information Bulletin, no. 728, p. 30.

4. Ibid., p. 19, table 5.

5. Ibid, p. 3.

6. John Gever, Robert Kaufmann, David Skole, and Charles Vorosmarty. *Beyond Oil* (Cambridge, MA: Ballinger Publishing, 1986), pp. 160-161.

7. Mary Myers, "Grass, Legumes, and a Reverence for Soil Fertility Allow Small Acreage Amish Farms to Thrive in Today's Economy," *The Stockman Grass Farmer 55*(6), (June 1998), pp. 1-2.

8. Gene Logsdon, "A Lesson for the Modern World," *Whole Earth Review,* (Spring 1986), pp. 81-82.

9. Stewart Smith, "Farming: It's Declining in the U.S." *Choices 7*(1), (First quarter 1992), pp. 8-10.

10. Seth M. Dabney, "Cover Crop Integration into Conservation Production Systems," *USDA Sustainable Agriculture Research and Education Program, Southern Region, 1997 Annual Report,* p. 11.

11. National Public Radio, *Morning Edition,* "Organic Farming Series, November 1, 1994 quoted in Tracy Irwin Hewitt and Katherine R. Smith, "Intensive Agriculture and Environmental Quality: Examining the Newest Agricultural Myth," Report from the Henry A. Wallace Institute for Alternative Agriculture (Greenbelt, MD, September 1998), p. 4.

12. Charles A. Griffith, "Why Rotational Stocking Through a Grazing Cell Is a Way to Improve Income and Ecological Soundness of a Ranch," *Pasture Management Clinic Proceedings* (Poteau, OK: Kerr Center for Sustainable Agriculture, 1994).

13. Karl Kessler, "Direct Marketing Booms," *The Furrow 103*(7), (September-October 1998), pp. 10-11.

14. Ibid.

15. "Smashing the Hourglass," *Land Stewardship Letter,* p. 13.

16. John Dietz, "Chasing the Carrot," *The Furrow 130*(5), (spring 1998), pp. 27-28.

17. Jules Janick, Melvin G. Blase, Duane L. Johnson, Gary D. Joliff, and Robert L. Meyers, "Diversifying U.S. Crop Production," *Council for Agricultural Science and Technology (CAST) Issue Paper #6* (Ames, IA: CAST, February 1996).

18. Del Deterling, "Fueling the Boom in Organics," *Progressive Farmer 113*(4), (March 1998) p. 56.

19. Organic Farming Research Foundation, *Organic Certifiers Directory,* Available online: <http://www.ofrf.org/about_organic/certifier.html>.

20. Julie Anton Dunn, *Organic Food and Fiber: An Analysis of 1994 Certified Production in the United States* (Washington, DC: Department of Agriculture Agricultural Marketing Service, 1995).

21. Department of Agriculture, National Organic Standards Board, "Definition of Organic," in Organic Farmers Marketing Association, Available online: <http://web.iquest.net/ofma/defin.htm>, p. 1. [January 1, 2001].

22. Deterling, "Fueling the Boom in Organics," p. 57.
23. Ibid., p. 54.
24. U.S. Department of Agriculture, Economic Research Service Natural Resources and Environment Division, *Updates of Agricultural Resources and Environmental Indicators,* AREI Updates No. 4, May 1996.
25. Ruckelshaus, "Toward a Sustainable World."
26. "Swine Production Costs," *Doane's Agricultural Report 60*(32), (August 8, 1997), p. 5.
27. Chris Ison, "State Health Department Acknowledges Health Risks of Feedlots," *Minneapolis Star Tribune 18*(322), (February 20, 2000), p. B-1.
28. Judith D. Soule and Jon K. Piper, *Farming in Nature's Image* (Washington, DC: Island Press, 1992), p. 13.
29. John Pesek, "Historical Perspective," in *Sustainable Agriculture Systems,* Eds. J.L. Hatfield and D.L. Karlen (Boca Raton, FL: Lewis Publishers, 1994), p. 6.
30. Ruckelshaus, "Toward a Sustainable World," p. 371.
31. R. Repetto, W. Magrath, C. Beer, and F. Rossini, *Wasting Assets: Natural Resources in the National Income Accounts* (Washington, DC: World Resources Institute, 1989) quoted in John De Boer, *Building Sustainable Agricultural Systems: Economic and Policy Dimensions,* Development Studies Paper Series (Morrilton, AR: Winrock International Institute for Agricultural Development, 1993), p. 18.
32. William Vorley, "The Pesticide Industry and Sustainable Agriculture," *Leopold Letter 6*(3), (Fall 1994).
33. Department of Agriculture, Economic Research Service, Food and Rural Economics Division, *Food Consumption, Prices and Expenditures, 1970-97,* by Judith Jones Putnam and Jane E. Allshouse, Statistical Bulletin no. 965 (Washington, DC, 1990), p. 14.

Chapter 10

1. Robert Frost, "The Road Not Taken," *The Poetry of Robert Frost,* Ed. Edward Connery Lathem (New York: Henry Holt, 1969), p. 131.
2. Earl May, "Cold Springs" in *Pioneering in Kiowa County,* (Hobart, OK: Kiowa County Historical Society, 1978).
3. David Strawn, "Rural Development Conference" (paper presented at the Kerr Center Rural Development Conference, Shawnee, OK, June 26, 1997).
4. U.S. Congress, Office of Technology Assessment, *Technology, Public Policy, and the Changing Structure of American Agriculture* (Washington, DC: GPO, 1986), p. 226 cited in Osha Gray Davidson, *Broken Heartland: The Rise of America's Rural Ghetto* (Iowa City: University of Iowa Press, 1996), pp. 166-167.
5. North Central Regional Center for Rural Development, *Bringing Home the Bacon: The Myth of the Role of Corporate Hog Farming in Rural Revitalization* (Poteau, OK: Kerr Center for Sustainable Agriculture, 1999), p. 3.
6. Ibid., p. ES 1-9.
7. Mitchell Satchell, "Hog Heaven and Hell: Pig Farming Has Gone High Tech and That's Causing New Polluting Woes," *U.S. News and World Report 120*(3), (January 22, 1996), p. 55.

8. M. Duncan, D. Fisher, and M. Drabenstatt, "America's Heartland: Can It Survive (Economic Growth in the Rural Great Plains)," *USA Today 14*(203), (July 1, 1996).

9. Walter Goldschmidt, *As You Sow* (Glencoe, IL: Free Press, 1947), quoted in Michael Perelman, "Efficiency in Agriculture," in *Radical Agriculture,* Ed. Richard Merrill (New York: Harper and Row, 1976), p. 82.

10. Robert E. Wagner, "Finding the Middle of the Road on Sustainability," *Journal of Production Agriculture 3,* (1990), pp. 277-280.

11. "Producer Involvement Boosts SARE Program," *USDA Sustainable Agriculture Research and Education Program (SARE) 1996 Project Highlights,* p. 2.

12. Tom Trantham, "It's About Time," in *1996 Southern Region Sustainable Agriculture Research and Education (SARE) Annual Report,* Ed. Gwen Roland (Griffin, GA: SARE, 1997), p. 1.

13. Thomas Jefferson, "Letter to John Wayles Eppes," 1813, in The University of Virginia, "Thomas Jefferson on Politics and Government," #38, p. 2, Available online: <http://etext.virginia.edu/jefferson/quotations/jeff1340.htm> [Dec. 14, 2000].

Bibliography

Agee, Chris. "Fungus, Grazing and Stocker Options." *Kerr Center for Sustainable Agriculture Newsletter 23*(3), p. 7.

Ambrose, Stephen E. *Undaunted Courage: Meriwether Lewis, Thomas Jefferson, and the Opening of the American West.* New York: Simon and Shuster, 1997, pp. 32, 33.

American Livestock Breeds Conservancy 1999 Breeders Directory, The. Pittsboro, NC: American Livestock Breeds Conservancy, 1999.

American Rivers. (n.d.). *America's Twenty Most Endangered Rivers of 1998.* Available online: <http://www.amrivers.org> [April 1999].

Anderson, John A. Jr., Peter T. Bromley, H. Michael Linker, Duane F. Newman, "Effects of Sustainable and Conventional Agriculture on Farm Wildlife," in *1994 Sustainable Agriculture Research and Education (SARE)/Agriculture in Concert with the Environment (ACE) Annual Report* Ed. Gwen Roland (Griffin, GA: SARE, 1995), p. 93.

Anderson, Julie. "States Monitor Hog Waste's Effects on Groundwater." *Omaha World Herald, 134*(72), December 28, 1998, p. C-6.

Anthan, George. "Seed Bank System May Wither." *Muskogee Phoenix and Times Democrat,* April 19, 1998, p. C-10.

Aspelin, A.L. *Pesticide Industry Sales and Usage: 1992 and 1993 Market Estimates.* U.S. Environmental Protection Agency. Washington, DC. June 1994. Quoted in Toni Nelson, "Efforts to Control Pesticides Expand," in *Vital Signs.* New York: W.W. Norton, 1996, p. 108.

Aspelin, A.L. *Pesticide Industry Sales and Usage: 1994 and 1995 Market Estimates—Preliminary.* Quoted in Charles M. Benbrook, Edward Groth III, Jean M. Holloran, Michael K. Hansen, and Sandra Marquardt, *Pest Management at the Crossroads.* Yonkers, NY: Consumers Union, 1996, p. 41.

Assadian, N. Waissman, L. Corral Esperanza, C. Ponce, "Organic Cotton Production—1993-1997," (Texas A&M Research Center, El Paso Texas, 1999, photocopy), p. 29.

Avery, Dennis T. *Saving the Planet with Pesticides and Plastics: The Environmental Triumph of High-Yield Farming.* Indianapolis, IN: Hudson Institute, 1995.

Bailey, Liberty Hyde. *The Holy Earth.* Lebanon, PA: Sowers Printing, 1915.

Benbrook, Charles M. Remarks to Sustainable Agriculture Scoping Group, President's Council on Sustainable Development, Washington, DC, April 28, 1994.

Benbrook, Charles M., Edward Groth III, Jean M. Holloran, Michael K. Hansen, and Sandra Marquardt. *Pest Management at the Crossroads.* Yonkers, NY: Consumers Union, 1996.

Berry, Wendell, "Essay." *Audubon, 95*(2), March/April 1993, pp. 102-106.

"Biotechnology," *Doane's Agricultural Report, 60*(37), September 12, 1997, p. 1.

"Biting Perspectives." California Sustainable Agriculture Research and Education Program Newsletter: *Inquiry in Action,* No. 21 (January 1999), p. 18.

Bixby, Donald E., Carolyn J. Christman, Cynthia J. Ehrman, and D. Phillip Sponenberg. *Taking Stock: The North American Livestock Census.* Blacksburg, VA: McDonald and Woodward Publishing, 1994.

Britton, Jim, Le Flore County Area Poultry Extension Specialist. Telephone conversation with Maura McDermott, April 1997.

Campbell, Colin J. and Jean H. Laherrere, "The End of Cheap Oil." *Scientific American, 278*(3), March 1998, pp. 78-83.

Cantrell, Jim. *A Low-Investment Swine System.* Poteau, OK: Kerr Foundation Agricultural Division, 1985.

Carson, Rachel. *Silent Spring.* Greenwich, CT: Fawcett, 1962.

Castro, Janine and Frank Reckendorf. "Effects of Sediment on the Aquatic Environment: Potential NRCS Actions to Improve Aquatic Habitats." Working Paper No. 6. Natural Resources Conservation Service. Oregon State University, Department of Geosciences, August 1995.

Center for Rural Affairs. "No-Till and Sustainability." *Beginning Farmer Newsletter,* March 1998, p. 3.

Cocannouer, Joseph A. *Farming with Nature.* Norman, OK: University of Oklahoma Press, 1954.

Comis, Don. "Farmers New Part of the Global Warming Solution as U.S. Agriculture Becomes Net Carbon Sink." Department of Agriculture, Agricultural Research Service News Service, May 17, 1999. Press release.

Council for Agricultural Science and Technology (CAST) (n.d.). *Benefits of Biodiversity.* Available online: <http://www.cast-science.org/biod/biod_ch.htm> [September 13, 1999].

Crummett, Dan. "Ridge-Till Coming to Oklahoma." *Oklahoma Farmer-Stockman, 105*(4), April 1992, pp. 5-7.

Crummett, Dan. "Can You Afford Not to No-Till?" *Oklahoma Farmer-Stockman, 12*(4), April 1999, pp. 8-9.

Cunningham, William P. and Barbara Woodworth Saigo. *Environmental Science—A Global Concern.* Dubuque, IA: William C Brown, 1990.

Cunningham, William P. and Barbara Woodworth Saigo. *Environmental Science—A Global Concern.* Third Edition. Dubuque, IA: William C Brown, 1995.

Dabney, Seth M. "Cover Crop Integration into Conservation Production Systems." *USDA Sustainable Agriculture Research and Education Program, Southern Region, 1997 Annual Report.*

Daniel, Pete. *Breaking the Land.* Chicago: University of Illinois Press, 1985.

Darnay, Arsen J., Ed. *Statistical Record of the Environment.* Detroit: Gale Research.

Darwin, Charles, Paul H. Barrett, and R.B. Freeman. *The Formation of Vegetable Mould Through the Action of Worms With Observations on Their Habits.* Reprint edition, Vol. 28. New York: New York University Press, 1990.

Department of Agriculture, National Organic Standards Board, "Definition of Organic," in Organic Farmers Marketing Association. Available online: <http://web.iquest.net/ofma/defin.htm>, p. 1 [January 1, 2001].

Department of Labor, Occupational Health and Safety Administration, "Agricultural Operations," p. 1, Available online: <http://www.osha.slc.gov/SLTC/agricultural operations> [January 2, 2001].

Department of Rural Sociology, "Concentration of Agricultural Markets." University of Missouri: October 1997. Photocopy.

Deterling, Del. "Fueling the Boom in Organics." *Progressive Farmer, 113*(4), March 1998, pp. 54-57.

Devlin, Jeanne M. "A Century of Oil," *Oklahoma Today, 47*(3), July-August 1997, pp. 59-63.

DeVore, Brian. "Biodiversity and Agriculture: A House Divided." *Land Stewardship Letter, 16*(5), November 1998, pp. 1, 7-9.

Dietz, John. "Chasing the Carrot." *The Furrow, 103*(5), Spring 1998, pp. 27-28.

DiGiacomo, Gigi and Harry Smith, "A Look at Who's Gobbling Up America's Food Dollar: Agribusiness Shortchanging Farmers and Consumers," *Farm Aid News, 3*(12), June 28, 1995, p. 1.

"Doane's Focus Report: Planting Decisions '97," *Doane's Agricultural Report, 60*(8), February 21, 1997, p. 5.

Duncan, M., D. Fisher, and M. Drabenstatt. "America's Heartland: Can It Survive (Economic Growth in the Rural Great Plains)." *USA Today, 14*(203), July 1, 1996, p. 3B.

Easterbrook, Gregg. "Good Pollution." *The New Yorker,* October 20/27, 1997, pp. 78-80.

Elliott, Luke and Judi Fox. *Solar Water Pumping*. Meadowcreek Technical Brief No. 2, Fox, AR: Meadowcreek, 1992.

"Endolsulfan Responsible for Alabama Fish Kill." Alabama Department of Environmental Management. August 24, 1995. Press release. From Pesticide Action network North America Updates Service. February 23, 1996. Quoted in Charles M. Benbrook, Edward Groth III, Jean M. Holloran, Michael K. Hansen, and Sandra Marquardt, *Pest Management at the Crossroads*. Yonkers, NY: Consumers Union, 1996, p. 64.

Environmental News Network, News Archive. August 30, 1995. *Study Finds High Nitrate Levels in Agricultural Areas*. Available online: <http://www.enn.com/enn-news-archive/1995/08/083095/08309519.asp> [February 17, 2000].

Environmental Working Group (n.d.). *Pouring It On: Nitrogen Use and Sources of Nitrate Contamination.* Available online: <http://www.ewg.org/pub/home/reports/Nitrate/NitrateUse.html> [July 1999].

Facts from Our Environment. Atlanta: Potash Institute of North America, 1972.

Fan, Anna M., Calvin C. Wilhite, and Steven A. Book. "Evaluation of the Nitrate Drinking Water Standard with Reference to Infant Methemoglobinemia and Potential Reproductive Toxicity." In *Regulatory Taxicology and Pharmacology* Volume 7 (1987) pp. 135-148. Quoted in Ann Ziebarth (n.d.). *Well Water, Nitrates and the 'Blue Baby' Syndrome Methemoglobinemia*. Available online: <http://ianrwww.unl.edu/IANR/PUBS/NEBFACTS/nf91-49.htm> [July 1999], p. 2.

Fanatico, Anne. "New Free Range Poultry Guidebook Covers Production, Processing and Promotion." *ATTRA News* 5(3), June 1997, p. 3.

"Farm Facts and Fancies." *The Furrow, 102*(8) November 1997, p. 40.

Fauci, Mary F. and R.P. Dick. "Soil Microbial Dynamics: Short- and Long-Term Effects of Inorganic and Organic Nitrogen," *Soil Science Society of America Journal, 58* (May-June 1994), pp. 801-806.

Faulkner, Edward H. *Plowman's Folly*. New York: Grosset and Dunlap, 1943.

"Fertilizers Growing a Dead Zone in Gulf." *Tulsa World*. January 25, 1999.

Filimonova, Maria. "Life Underground." *Kerr Center Newsletter* (November/December 1996), p. 9.

Food and Agriculture Organization (FAO) of the United Nations. *State of the World's Genetic Resources for Food and Agriculture*. Rome, 1996. Quoted in Hope Shand, *Human Nature: Agricultural Biodiversity and Farm-Based Food Security*. Ottawa: Rural Advancement Foundation International, 1997, p. 21.

Ford, Brian. "Lake Eucha Quality Talk Grim." *Tulsa World, 92*(342), August 21, 1997, pp. A-1, 4.

Francis, Charles and Garth Youngberg. "Sustainable Agriculture—an Overview." In Charles A. Francis, Cornelia Butler Flora, and Larry D. King (Eds.), *Sustainable Agriculture in Temperate Zones*. New York: Wiley, 1990, pp. 1-23.

Frost, Robert. "Mowing." In Edward Connery Lathem (Ed.), *The Poetry of Robert Frost*. New York: Henry Holt, 1969, p. 25.

Frost, Robert. "The Road Not Taken." In Edward Connery Lathem (Ed.), *The Poetry of Robert Frost*. New York: Henry Holt, 1969, p. 131.

Gardner, Gary. "Shrinking Fields: Cropland Loss in a World of Eight Billion." Worldwatch Paper 131, July 1996.

Genetic Resources Action International (GRAIN) (n.d.). *Intellectual Property Rights and Biodiversity: The Economic Myths*. Available online: <http://www.grain.org/publications/gtbc/issue3.htm> [August 1999]. Quoted in Manjula V. Guru and James E. Horne, *Biotechnology: A Boon or a Curse*. Poteau, OK: Kerr Center, January 2000, p. 4.

Gever, John, Robert Kaufmann, David Skole, and Charles Vorosmarty. *Beyond Oil*. Cambridge, MA: Ballinger Publishing, 1986.

Goldschmidt, Walter. *As You Sow*. Glencoe, IL: Free Press, 1947. Quoted in Michael Perelman, "Efficiency in Agriculture." In Richard Merrill (Ed.,) *Radical Agriculture*. New York: Harper and Row, 1976, p. 82.

Griffith, Charles A. "Why Rotational Stocking Through a Grazing Cell Is a Way to Improve Income and Ecological Soundness of a Ranch." *Pasture Management Clinic Proceedings*. Poteau, OK: Kerr Center for Sustainable Agriculture, 1994, pp. 19-26.

Guru, Manjula V. and James E. Horne. *Biotechnology: A Boon or a Curse*. Poteau, OK: Kerr Center, January 2000.

Hall, T. "In Cuisine of the 90's, the Farmer Is the Star," *The New York Times*, July 8, 1992. Quoted in Janet Bachman, "Sustainable Agriculture and Rural and Urban Communities: What Are the Connections?" *KCSA Newsletter 18*, November 1992, pp. 2-3.

Hamilton, Neil D. "Agriculture Without Farmers? Is Industrialization Restructuring American Food Production and Threatening the Future of Agriculture?" *Northern Illinois University Law Review, 14*(3), summer 1994, p. 630.

Hanson, Victor Davis. *Fields Without Dreams*. New York: The Free Press, 1996.

Harwood, Richard R. "The History of Sustainable Agriculture." In Clive A. Edwards, Rattan Lal, Patrick Madden, Robert H. Miller, and Gar House (Eds.), *Sustainable Agricultural Systems*. Delray Beach, FL: St. Lucie Press, 1990, pp. 3-19.

Hassebrook, Chuck. "Is There a Place for Biotechnology?" Paper presented at Future Farms conference, Oklahoma City, OK, February 8, 2000.

Hecht, Susanna B. "The Evolution of Agroecological Thought." In Miguel A. Altieri (Ed.), *Agroecology: The Scientific Basis of Alternative Agriculture*. Boulder, CO: Westview Press, 1987, pp. 1-20.

Hewitt, Tracy Irwin and Katherine R. Smith. "Intensive Agriculture and Environmental Quality: Examining the Newest Agricultural Myth." *Report from the Henry A. Wallace Institute for Alternative Agriculture*. Greenbelt, MD: September 1998.

Hill, S.B. and Jacques Nault. "Towards a More Energy Efficient Cash Crop Farm." *Sustainable Farming, 6*(1), Fall 1995, pp. 8-11.

Holdren, John P. "Energy Agenda for the 1990's." *Energy Use Trends*. In Jack M. Hollander (Ed.), *The Energy-Environment Connection*. Washington, DC: Island Press, 1992, pp. 377-391.

Hollander, Jack M. Jack M. Hollander (Ed.), Introduction to *The Energy-Environment Connection*. Washington, DC: Island Press, 1992.

House Appropriations Subcommittee on Agriculture, Rural Development, and Related Agencies. Testimony of Jess Andrews III. *Low Input Agricultural Research, Hearings on Subsection C of Title 14 of the 1985 Farm Bill*. Washington, DC, April 8, 1987.

House Appropriations Subcommittee on Agriculture, Rural Development, and Related Agencies. Testimony of Brian Chabot. *Low Input Agricultural Research, Hearings on Subsection C of Title 14 of the 1985 Farm Bill*. Washington, DC, April 8, 1987.

House Appropriates Subcommittee on Agriculture, Rural Development, and Related Agencies. Testimony of Charles A. Francis. *Low Input Agricultural Research Hearings on Subsection C of Title 14 of the 1985 Farm Bill*. Washington, DC, April 8, 1987.

Howard, Sir Albert. *An Agricultural Testament*. New York: Oxford University, 1943.

Hurt, R. Douglas. *American Agriculture: A Brief History*. Ames, IA: Iowa State University Press, 1994.

IMC Global, Inc. (n.d.). *World Crop Nutrients and Salt Situation*. Available online: <http://www.imcglobal.com/cropsalt/k20rgncap.htm> [May 1999].

IMC Global, Inc. (n.d.). *World Crop Nutrients and Salt Situation*. Available online: <http://www.imcglobal.com/cropsalt/nak20cap.htm> [May 1999].

Ikerd, John E. "Sustainable Agriculture: Farming in Harmony with the Biosphere." In *Sustainable Agriculture: Enhancing the Environmental Quality of the Tennessee Valley Region through Alternative Farming Practices*. University of Tennessee Extension Service, EC 1022-750-9/92. Quoted in Janet Bachman. "Sustainable Agriculture and Rural and Urban Communities: What are the Connections." *KCSA Newsletter, 18*(11), November 1992, pp. 2-3.

Ison, Chris. "State Health Department Acknowledges Health Risks of Feedlots." Minneapolis *Star Tribune, 18*(322) February 20, 2000, p. B-1.

Jacobs, T.C. and J.W. Gilliam. "Riparian Losses of Nitrate from Agriculture Drainage Waste." *Journal of Environmental Quality 14*(1985). Quoted in Oklahoma Cooperative Extension Service, Oklahoma State University, and Oklahoma Conservation Commission. *Riparian Area Management Handbook.* Stillwater, OK: Oklahoma State University, 1998.

Janick, Jules, Melvin G. Blasé, Duane L. Johnson, Gary D. Joliff, and Robert L. Meyers, "Diversifying U.S. Crop Production," *Council for Agricultural Science and Technology (CAST) Issue Paper #6.* Ames, IA: CAST, February 1996.

Jefferson, Thomas. "Letter to John Wayles Eppes," 1813, in The University of Virginia, "Thomas Jefferson on Politics and Government," #38, p. 2, Available online: <http://etext.virginia.edu/jefferson/quotations/jeff1340.htm> [December 14, 2000].

Jennings, Vivian M. and John Ikerd. "National Perspective on Agricultural Sustainability." Paper presented at the American Society of Agronomy North Central Branch Meeting in conjunction with the Registry for Environmental and Agricultural Professional (REAP) Workshop. Des Moines, IA, August 1, 1994.

Jeyaratnam, J. "Acute Pesticide Poisoning: A Major Problem." *World Health Statistics Quarterly, 43*(3) (1990). Quoted in Lester R. Brown, Christopher Flavin, and Hal Kane, *Vital Signs 1996.* New York: W.W. Norton, 1996, pp. 108-109.

Jobes, Allen. "Crop Rotation for Disease and Insect Control." *Kerr Center Newsletter, 23*(1), 1997, p. 2.

Kaplan, J. Kim. "Conserving the World's Plants." *Agricultural Research.* September 1998, p. 8.

Kennedy, Robert F. Jr. "I Don't Like Green Eggs and Ham!" *Newsweek,* April 26, 1999, p. 12.

Kenney, Dennis R. "Biological Control: Its Role in Sustainable Agriculture." *Leopold Letter, 8*(4), winter 1996, pp. 3, 6.

Kerr, Robert S. *Land, Wood, and Water.* New York: Fleet Publishing Corporation, 1960.

Kessler, Karl. "Direct Marketing Booms." *The Furrow, 103*(7), September-October 1998, pp. 10-13.

Kidwell, Boyd with Earl Manning and Karl Wolfshohl. "The Big Push for Buffer Strips." *Progressive Farmer, 113*(12), November 1998, pp. 20-23.

"Kinder Killers: Many Little Hammers." *The American Gardener,* March/April 1997.

Kirschenmann, Frederick. *Switching to a Sustainable System.* Windsor, ND: Northern Plains Sustainable Agriculture Society, 1998.

Kleinschmit, Martin. "Putting Life Back into the Soil," *Beginning Farmer Newsletter,* June 1999, p. 4.

Lathrop, Will. "The Kerr Center Ranch, 1987-1996." Kerr Center for Sustainable Agriculture, Poteau, OK. Photocopy.

Le Flore County Conservation District. "Long Range Total Resource Conservation Program July 1994-June 1999." Le Flore County Conservation District, Poteau, OK. Photocopy.

Lewis, W. Joe, J.C. van Lenteren, Sharad C. Phatak, and J.H. Tumlinson III, "A Total System Approach to Sustainable Pest Management." In *Proceedings of The National Academy of Science 94,* (November 1997), pp. 1223-1248.

Logsdon, Gene. "A Lesson for the Modern World." *Whole Earth Review,* Spring 1986, pp. 81-82.

Lovins, Amory B., L. Hunter Lovins, and Marty Bender, "Energy and Agriculture." In Wes Jackson, Wendell Berry, and Bruce Colman (Eds.), *Meeting the Expectations of the Land.* San Francisco: North Point Press, 1984, pp. 68-86.

"Low-Input Farming Systems: Benefits and Barriers." *Seventy-Fourth Report by the Committee of Government Operations.* Washington, DC: GPO, 1986.

MacCannell, D. "Agribusiness and the Small Community," background paper to *Technology, Public Policy and the Changing Structure of American Agriculture.* Office of Technology Assessment. Washington, DC: GPO, 1983. Quoted in Janet Bacman. "Sustainable Agriculture and Rural and Urban Communities: What Are the Connections?" *KCSA Newsletter 18*(11), November 1992, pp. 2-3.

Madden, J. Patrick. "Goals and Realities of Attaining a More Sustainable Agriculture." Paper presented at the Environmentally Sound Agriculture conference, Orlando, FL, April 1994.

May, Earl. "Cold Springs." In *Pioneering in Kiowa County.* Hobart, OK: Kiowa County Historical Society, 1978, pp. 223-231.

McMillen, Wheeler. *Feeding Multitudes.* Danville, IL: Interstate Printers and Publishers, 1982.

McPhee, John. "Encounters with the Archdruid" (1971). Quoted in Barbara K. Rodes and Rice Odell. *A Dictionary of Environmental Quotations.* New York: Simon and Schuster, 1992, pp. 85-86.

Moulton, Mark and Nancy Moulton (n.d.). "Deep Straw in Hoops: Managing Manure in Concert with the Natural, Social and Economic Environment." *Purdue University Manure Management* Paper 37. Available online: <http://kyw.ctic.purdue.edu/FRM/ManureMgmt/Paper37.html> [July 1999].

Myers, Mary. "Grass, Legumes, and a Reverence for Soil Fertility Allow Small Acreage Amish Farms to Thrive in Today's Economy." *The Stockman Grass Farmer, 55*(6), June 1998, pp. 1-2.

National Campaign for Sustainable Agriculture. *Environmental Degradation and Public Health Threats from Factory Farm Pollution.* Fact Sheet No. 1. November 17, 1998.

National Campaign for Sustainable Agriculture. *Sustainable Alternatives to Factory Farm Animal Production.* Fact Sheet No. 2, November 1998.

National Public Radio, *Morning Edition,* "Organic Farming Series, November 1, 1994. Quoted in Tracy Irwin Hewitt and Katherine R. Smith. *Intensive Agriculture and Environmental Quality: Examining the Newest Agricultural Myth.* Greenbelt, MD: Henry A. Wallace Institute for Alternative Agriculture, 1995, p. 9.

National Research Council. *Report on Alternative Agriculture.* Washington, DC: National Academy Press, 1989.

New, Leon (n.d.). *Opportunities to Maximize the Utilization of Water by Irrigators.* P&R Surge Systems. Available online: <http://www.prsurge.com/otmtuowb.htm> [July 1999].

"Newswatch." *Progressive Farmer, 112*(2), December 1997, p. 6.

Nichols, Mike. "Agriculture Observes Earth Day Every Day," (Oklahoma Farm Bureau, Oklahoma City, Oklahoma, April 21, 1998, press release), p. 1.

Norby, Monica Manton. "Improved Cropping Systems to Protect Groundwater Quality." *The Aquifer, 13*(2), August 1998, p. 6.

North Central Regional Center for Rural Development. *Bringing Home the Bacon: The Myth of the Role of Corporate Hog Farming in Rural Revitalization.* Poteau, OK: Kerr Center for Sustainable Agriculture, 1999.

Oklahoma Biodiversity Task Force, Norman L. Murray (Ed.), *Oklahoma's Biodiversity Plan: A Shared Vision for Conserving Our Natural Heritage.* Oklahoma City: Oklahoma Department of Wildlife Conservation, 1996. Draft copy.

Oklahoma Conservation Commission. *Poteau River Comprehensive Watershed Management Program.* Oklahoma FY-94, 319h work plan. March 1995.

Oklahoma State University. Department of Agricultural Economics. "Bottomland Soybean Budget for Southeast Oklahoma." March 15, 1994.

Oklahoma Water Resources Board. *Synopsis of the Oklahoma Comprehensive Water Plan.* Pub. 94-S, January 1980.

Oklahoma Water Resources Board. Water Quality Programs Division. *Diagnostic and Feasibility Study of Wister Lake, Executive Summary.* Oklahoma City: Oklahoma Water Resources Board, May 1, 1996.

Oliver, Charles. "Paying U.S. Farmers to Pollute: Subsidies Spur Increased Pesticide, Fertilizer Use." *Investors Business Daily—National Issue,* October 6, 1995, pp. A5-A6.

Olkowski, Helga and William Olkowski. "Design of Small-Scale Food Production Systems and Pest Control." In Dietrich Knorr (Ed.), *Sustainable Food Systems.* Westport, CT: AVI Publishing Company, 1983, pp. 143-155.

Organic Farming Research Foundation, *Organic Certifiers Directory,* Available online: <http://www.ofrf.org/about_organic/certifier.html>.

Parr, J.F., R.I. Papendick, I.G. Youngberg, and R.E. Meyer, "Sustainable Agriculture in the United States." In Clive A. Edwards, Rattan Lal, Patrick Madden, Robert H. Miller, and Gar House (Eds.), *Sustainable Agricultural Systems.* Delray Beach, FL: St. Lucie Press, 1990, pp. 50-67.

Paul James. "Internalizing Externalities: Granular Carbofuran Use on Rapeseed in Canada." *Ecological Economics 13*(3), June 1995, pp. 181-184. Quoted in Tracy Irwin Hewitt and Katherine R. Smith. "Intensive Agriculture and Environmental Quality: Examining the Newest Agricultural Myth." *Report from the Henry A. Wallace Institute for Alternative Agriculture.* Greenbelt, MD, September 1998, p. 4.

Perelman, Michael. "Efficiency in Agriculture." In Richard Merrill (Ed.), *Radical Agriculture.* New York: Harper and Row, 1976, pp. 64-86.

Pesek, John. "Historical Perspective." In J.L. Hatfield and D.L. Karlen (Eds.), *Sustainable Agriculture Systems.* Boca Raton, FL: Lewis Publishers, 1994, pp. 1-19.

Pesticide Action Network of North America (n.d.). *Exports of Hazardous Pesticides from U.S. Ports Increases.* Available online: <http://www.panna.org/panna/>. [May 26, 1998].

Phillips, F. Dwain. "Potential for Wind Erosion Is at 20 Year High." U.S. Department of Agriculture, National Resources Conservation Service. Stillwater, OK, March 12, 1996. Press release.

"Producer Involvement Boosts SARE Program." *USDA Sustainable Agriculture Research and Education Program (SARE) 1996 Project Highlights.*

Repetto, Robert and Sanjay S. Baliga. *Executive Summary: Pesticides and the Immune System: The Public Health Risks.* Washington, DC: World Resources Institute, 1996.

Repetto, Robert and Sanjay S. Baliga. *Pesticides and the Immune System: The Public Health Risks.* Washington, DC: World Resources Institute, 1996.

Repetto, R., W. Magrath, C. Beer, and F. Rassini. *Wasting Assets: Natural Resources in the National Income Accounts.* Washington, DC: World Resources Institute, 1989. Quoted in John De Boer, *Building Sustainable Agricultural Systems: Economic and Policy Dimensions.* Development Studies Paper Series. Morrilton, AR: Winrock International Institute for Agricultural Development, 1993, p. 18.

Repogle, Kenneth. "Sustainable Family Farming," (Remarks to the President's Council on Sustainable Development, Coweta, Oklahoma, September 23, 1997).

Rhoades, Robert. "The Incredible Potato." *National Geographic,* May 1982. Quoted in Phil Williams. "A Man for All Seasons (and Time Zones)." *Georgia Alumni Record, 73*(1), December 1993, p. 17.

Roberts, Cokie and Steven V. Roberts. "Hunger: A Startling Crisis." *USA Weekend,* March 27-29, 1998, pp. 4-7.

Roosevelt, Theodore. December 8, 1908 quoted in Harold Evans, "Saving America the Beautiful," in Random House, *The American Century, Old and New Americans 1880-1910,* p. 2. Available online: <http://www.randomhouse.com/features/american century/oldnew.html> [December 2000].

Ruckelshaus, William D. "Toward a Sustainable World." In *The Energy-Environment Connection,* edited by Jack M. Hollander. Washington, DC: Island Press, 1992, pp. 365-382.

Saign, Geoffrey C. *Green Essentials.* San Francisco: Mercury House, 1994.

Salatin, Joel. Introduction to *Pastured Poultry Profits.* Swoope, VA: Polyface, Inc., 1993, pp. v-vii.

Satchell, Mitchell. "Hog Heaven and Hell: Pig Farming Has Gone High Tech and That's Causing New Polluting Woes." *U.S. News and World Report, 120*(3), January 22, 1996, p. 55.

Schafer, Shaun. "Erosion Program Enrollment Blowin' in the Wind." *Tulsa World, 92*(14), September 25, 1996, p. 3.

Schaller, Neill. "Federal Policies to Fully Support Sustainable Agriculture Research and Education." In A. Ann Sorenson (Ed.), *Agricultural Conservation Alternatives: The Greening of the Farm Bill.* DeKalb, IL: American Farmland Trust Center for Agriculture in the Environment, 1994, pp. 72-82.

Science and Technology News Network (n.d.). *Killer Corn—Local Tips.* Available online: <http://www.stm2.net/pages11/killercorn/ltips.html> [September 7, 1999].

Scrinis, Gyorgy (n.d.). Colonizing the Seed. *Canberra Organic Growers Society, Inc.* Available online: <http://www.netspeed.com/au/cogs/gen3.htm> [August, 1999]. Quoted in Manjula V. Guru and James E. Horne. *Biotechnology: A Boon or a Curse.* Poteau, OK: Kerr Center, January 2000, p. 4.

Shand, Hope. *Human Nature: Agricultural Biodiversity and Farm-Based Security.* Ottawa: Rural Advancement Foundation International, 1997.

SimpleLife (n.d.). *Organic Cotton Exhibit.* Available online: <http://www.simplelife.com/organicctton/001-ORGANIC%20COT.html> [May 19, 1998].

Sinclair, Ward. "How's That Again, Mr. Deputy Secretary?" *The Washington Post,* April 29, 1988, p. A-5.

"Smashing the Hourglass." *Land Stewardship Letter, 17*(1), January-March 1999, p. 13.

Smith, Bruce D. *The Emergency of Agriculture.* New York: Scientific American Library, 1995.

Smith, Stewart. "Farming: It's Declining in the U.S." *Choices, 7*(1), First quarter 1992, pp. 8-10.

Snapp, Roscoe R. and A.L. Newmann. *Beef Cattle.* Fifth Edition. New York: John Wiley and Sons Inc., 1996.

Sobocinski, Paul. "The Crisis That Monocropping Built." *Land Stewardship Letter, 16*(5), November 1998, pp. 2-3.

Soule, Judith D. and Jon K. Piper. *Farming in Nature's Image.* Washington, DC: Island Press, 1992.

Staff, wire reports. "State Resists Conservation Tillage." *Tulsa World, 93*(46), October 28, 1997, p. E-3.

Stanley, Doris. "Testing Diversified Orchard Ecosystems: Cover Crops, Mixed Trees Encourage Beneficial Insects." *Agricultural Research, 44*(1), January 1996, pp. 18-19.

Strange, Marty. *Family Farming: A New Economic Vision.* Lincoln, NE: University of Nebraska Press, 1988.

Strange, Marty, Liz Krupicka, and Dan Looker. "The Hidden Health Effects of Pesticides." In *It's Not All Sunshine and Fresh Air: Chronic Health Effects of Modern Farming Practices.* Walthill, NE: Center for Rural Affairs, 1984, pp. 55-75.

Strawn, David. "Rural Development Conference." Paper presented at the Kerr Center Rural Development Conference. Shawnee, OK, June 26, 1997.

Suskziw, Jan. "Biological Warfare Against Beet Armyworms." *Agricultural Research, 46*(1) January 1998, p. 17.

"SWCS Policy Position Statement: Sustainable Agriculture," *Journal of Soil and Water Conservation, 50*(6), November/December 1995, p. 635.

"Swine Production Costs." *Doane's Agricultural Report, 60*(32), August 8, 1997, p. 5.

Tangley, Laura. "How to Create a Life Without Sex." *U.S. News and World Report, 127*(4), July 26, 1999, p. 41.

Texas Organic Marketing Cooperative "Mission Statement," Texas Organic Marketing Cooperative, O'Donnell, Texas, 1998, photocopy, p. 2.

Thompson, L.M. and F.R. Troch. *Soil and Soil Fertility.* New York: Oxford University, 1978.

Trantham, Tom. "It's About Time." In *Sustainable Agriculture Research and Education Program (SARE) Southern Region 1996 Annual Report,* edited by Gwen Roland. Griffin, GA., 1997, p. 1.

Trewe, Lynne. "The Other Endangered Species." *Permaculture Drylands Journal, 28* (spring 1997), pp. 17-19.

U.N. Food and Agricultural Organization. *Estimated Energy Intensiveness of Food Crops.* 1987. Quoted in William P. Cunningham and Barbara Woodworth Saigo.

Environmental Science, Third Edition. Dubuque, IA: William C Brown, 1995, p. 231, table 11.4.

U.S. Bureau of the Census. "Farm Income and Expenses: 1980-1995." *Statistical Abstract of the United States 1997,* 117 Edition. Washington, DC, 1997.

U.S. Congress. Office of Technology Assessment. *Technology, Public Policy, and the Changing Structure of American Agriculture.* Washington, DC: GPO, 1986. In Osha Gray Davidson, *Broken Heartland: The Rise of America's Rural Ghetto.* Iowa City: University of Iowa Press, 1996, pp. 166-167.

U.S. Department of Agriculture. Agricultural Marketing Service. "Organic Food and Fiber: An Analysis of 1994 Certified Production in the United States." By Julie Anton Dunn. 1995.

U.S. Department of Agriculture. Economic Research Service. *Agricultural Resources and Environmental Indicators.* Agricultural Handbook no. 705. Quoted in Charles M. Benbrook, Edward Groth III, Jean M. Holloran, Michael K. Hansen, and Sandra Marquardt, *Pest Management at the Crossroads.* Yonkers, NY: Consumers Union, 1996.

U.S. Department of Agriculture. Economic Research Service. Agriculture and Rural Economy Division. *Economic Indicators of the Farm Sector: National Financial Summary 1992. ECFIC 12-1,* calculated from table 49. Quoted in Patricia Allen, *The Human Face of Sustainable Agriculture.* University of California, Santa Cruz, Sustainability in the Balance Series, Issue Paper No. 4, 1994, p. 4.

U.S. Department of Agriculture. Economic Research Service. Food and Rural Economics Division. *Food Consumption, Prices and Expenditures, 1970-97,* by Judith Jones Putnam and Jane E. Allshouse. Statistical Bulletin no. 965. Washington, DC, 1990.

U.S. Department of Agriculture. Economic Research Service. Natural Resources and Environment Division. *Agricultural Research.* AREI Updates No. 5 Revised. 1995.

U.S. Department of Agriculture. Economic Research Service. Natural Resources and Environment Division. *1995 Nutrient Use and Practices on Major Field Crops.* AREI Updates No. 2. May 1996.

U.S. Department of Agriculture. Economic Research Service. Natural Resources and Environment Division. *Pest Management on Major Field Crops.* AREI Updates No. 1, February 1997.

U.S. Department of Agriculture. Economic Research Service. Natural Resources and Environment Division. *Updates of Agricultural Resources and Environmental Indicators.* AREI Updates No. 4, May 1996.

U.S. Department of Agriculture. Natural Resources Conservation Service. *A Geography of Hope.* Program Aid 1548. 1996.

U.S. Department of Agriculture. Natural Resources Conservation Service. *Water Quality.* RCA Issue Brief 9. March 1996.

U.S. Department of Agriculture. Natural Resources Conservation Service. Soil Quality Institute. *Effects of Soil Erosion on Soil Productivity and Soil Quality.* Soil Quality-Agronomy Technical Note No. 7. August 1998.

U.S. Department of Agriculture. Natural Resources Service (April 1995). *Natural Resources Inventory: A Summary of Natural Resource Trends in the U.S. Between 1982 and 1992*. Available online: (nttp://www.greatplains.org/resource/nrisumm/LONESOIL.HTM> [March 1999].

U.S. Department of Agriculture. Natural Resources Service (April 1995). *Natural Resources Inventory: A Summary of Natural Resource Trends in the U.S. Between 1982 and 1992*. Available online: <http://greatplains.org/resource/nrisumm/NONFED.HTM> [March 1999].

U.S. Department of Agriculture. Natural Resources Service (April 1995). *Natural Resources Inventory: A Summary of Natural Resource Trends in the U.S. Between 1982 and 1992*. Available online: <http://greatplains.org/resource/nrisumm/SLIDE.HTM> [March 1999].

U.S. Department of Agriculture. Soil Conservation Service. *Farming with Residues*. Des Moines, IA, 1991.

U.S. Department of Agriculture. Sustainable Agriculture Research and Education Program, North Central Region. *1997 Annual Report*. Lincoln, NE, 1998.

U.S. Department of Commerce. National Technical Information Service. Economic Research Service. *Soil Erosion and Conservation in the United States: An Overview*. AIB718. October 1995.

U.S. Department of the Interior. Fish and Wildlife Service. *Why Save Endangered Species*. Brochure.

U.S. Department of Interior. U.S. Geological Survey Circular 1090. *Persistence of the DDT Pesticide in the Yakima River Basin Washington*. Washington, DC: GPO, 1993.

U.S. Environmental Protection Agency. *National Water Quality Inventory*. Report to Congress (Washington, DC, 1990). Quoted in Arsen Darnay, Ed., *Statistical Record of the Environment*. Detroit: Gale Research, Inc.

U.S. Environmental Protection Agency. *National Water Quality Survey* (1994). Quoted in U.S. Department of Agriculture. Natural Resources Conservation Service. *Water Quality*. RCA Issue Brief 9. March 1996.

U.S. Environmental Protection Agency. *Pesticides in Ground Water: Background Document* WH- 550G. Washington, DC: GPO, 1986. Quoted in Jim Bender. *Future Harvest*. Lincoln: University of Nebraska Press, 1994, p. 112.

U.S. General Accounting Office. *Sustainable Agriculture: Program Management, Accomplishments and Opportunities*. GAO/RCEDS-92-233.

U.S. Water News Online. (April 1996). *High Plains Drought Endangers Agriculture, Ogallala Aquifer*. Available online: <http://www.uswaternews.com/archive/96/supply/ogallala.htm> [January 1999].

United States Department of Agriculture Study Team on Organic Farming. *Report and Recommendations on Organic Farming*. July 1980.

University of California Sustainable Agriculture Research and Education Program. (n.d.). *What is Sustainable Agriculture*. Available online: <http://www.sarep.ucdavis.edu/sarep/concept.html#Natural Resources> [November 6, 1997].

"Up and Down Wall Street—Crude Awakening." *Barron's*, April 7, 1996. Quoted in "Smooth Sailing or Rough Seas Ahead for Energy Prices?" Royale Energy Report (May 1996). Available online: <http://www.royl.com/nl/5-96/html#GEOLOGY> [January 1997], pp. 1-2.

Vogelsang, Donald L. *Local Cooperatives in Integrated Pest Management*. U.S. Department of Agriculture Farmer Cooperative Service Research Report 37, February 1977.

Vorley, William. "The Pesticide Industry and Sustainable Agriculture." *Leopold Letter, 6*(3) (fall 1994), pp. 4-5.

Wagner, Robert E. "Finding the Middle of the Road on Sustainability." *Journal of Production Agriculture, 3* (1990), pp. 277-280.

Warner, R. E. "Illinois Farm Programs: Long Term Impacts on Terrestrial Ecosystems and Wildlife-Related Recreation, Tourism and Economic Development." *Report prepared for the Research and Planning Division*. Illinois Department of Energy and Natural Resources. Springfield, IL, 1991. Quoted in Tracy Irwin Hewitt and Katherine R. Smith. "Intensive Agriculture and Environmental Quality: Examining the Newest Agricultural Myth." *Report from the Henry A. Wallace Institute for Alternative Agriculture*. Greenbelt, MD, September 1998, p. 4.

Welsh, Rick. *Reorganizing U.S. Agriculture: The Rise of Industrial Agriculture and Direct Marketing*. Greenbelt, MD: Henry A. Wallace Institute for Alternative Agriculture, 1997.

Williams, Gregory and Patricia Y. Williams, "Nutrition Affects Susceptibility of Corn to Borer Damage." *HORTIDEAS, 15*(8), (August 1998), p. 89.

Williams, Kenneth. "Five Reasons Why You Should Consider Sustainable Farming Practices." *KCSA Newsletter, 17*(3), March 1991, pp. 2-3.

Willis, Harold. *The Coming Revolution in Agriculture*. Wisconsin Dells, 1985.

World Health Organization/UNEP. *The Public Health Impact of Pesticides Used in Agriculture*. Geneva: 1990. Quoted in Lester R. Brown, Christopher Flavin, and Hal Kane, *Vital Signs 1996*. New York: W.W. Norton, 1996, pp. 108-109.

Worster, Donald. *Nature's Economy: A History of Ecological Ideas*. Cambridge, UK: Cambridge University Press, 1977.

Zahm, Sheila Hoar and Aaron Blair. "Pesticides and Non-Hodgkins Lymphoma," *Cancer Research, 52* (October 1, 992): 5485s-5488s, quoted in Tracy Irwin Hewitt and Katherine R. Smith, "Intensive Agriculture and Environmental Quality: Examining the Newest Agricultural Myth," *Report from the Henry A. Wallace Institute for Alternative Agriculture* (Greenbelt, MD, September 1998), p. 5.

Ziebarth, Ann (n.d.). *Well Water, Nitrates and the 'Blue Baby' Syndrome Methemoglobinemia*. Available online: <http://ianrwww.unl.edu/IANR/PUBS/NEBFACTS/nf91-49.htm> [July 1999].

Index